NET-SLAVES 2.0

Tales of "Surviving" the Great Tech Gold Rush

**Bill Lessard
and Steve Baldwin**

Authors of NetSlaves®: True Tales of Working the Web

ALLWORTH PRESS
NEW YORK

© 2003 Bill Lessard and Steve Baldwin

All rights reserved. Copyright under Berne Copyright Convention, Universal Copyright Convention, and Pan-American Copyright Convention. No part of this book may be reproduced, stored in a retrieval system, or transmitted in any form, or by any means, electronic, mechanical, photocopying, recording, or otherwise, without prior permission of the publisher.

08 07 06 05 04 03   5 4 3 2 1

Published by Allworth Press
An imprint of Allworth Communications, Inc.
10 East 23rd Street, New York, NY 10010

Cover and interior design by James Victore, New York, NY
Cover photo by Teri Baldwin
Page composition/typography by Integra Software Services Pvt Ltd., Pondicherry, India

ISBN: 1-58115-284-1

Library of Congress Cataloging-in-Publication Data:

Lessard, Bill.
  Netslaves 2.0 : tales of "surviving" the great tech gold rush / Bill Lessard, Steve Baldwin.
    p. cm.
  ISBN 1-58115-284-1
  1. High technology industries—United States—Employees. 2. United States—Economic conditions—1981-2001. 3. United States—Economic conditions—2001– I. Title: Netslaves Two.0. II. Baldwin, Steve, 1956 July 10– III. Title.

HD8039.H542U65 2003
331.7'61004678'0973—dc21
2002155807

Printed in Canada

Table of Contents

v / **Acknowledgements**

vii / **Introduction**
Created on Acid, Built on Caffeine, Died on Prozac

1 / **Chapter 1**
Neo-Luddites: Take This Industry and Shove It!

25 / **Chapter 2**
Panhandlers: Will Code for Food

45 / **Chapter 3**
Vigilantes: Screaming for Justice

71 / **Chapter 4**
Shape-Shifters: See Them Change

97 / **Chapter 5**
Aliens: Twice the Work for Half the Pay

121 / **Chapter 6**
Pawns: Twenty Years of Typing and They Put You on the Web Shift

147 / **Chapter 7**
Bootstrappers: Forty Megs and a Stool

167 / **Chapter 8**
Grave Robbers: Picking the Bones Clean

189 / **Chapter 9**
Lepers: Are We Not Men?

207 / **Afterword**
What Next?

209 / **Index**

Acknowledgements

We would like to thank everyone who supported and believed in the NetSlaves project. In particular, we would like to thank:

The good folks at Allworth Press, for the fine work they did in editing this text.

Teri Baldwin, for her brilliant photographic skills and for the courage she displayed in "temporarily borrowing" the garbage can on the cover from the city of Yonkers.

Steve Gilliard and Patrick Neeman, for keeping NetSlaves.com continually stocked with new online content and improved functionality.

Family and friends, including: Teri Baldwin, Teri Hennelly, Tess Baldwin, Carl and Mary Ellen Baldwin, and Julie Hanlon.

And finally, we owe our deepest gratitude to the men and women who stepped forward and told us their stories. Without you, this work would have been impossible.

Introduction

— <u>Created on Acid, Built on Caffeine, Died on Prozac</u> —

The Great Recession—now in its third year—began with the dot.com crash. Those people working in what used to be called "IT," but is now referred to simply as "that there idiot with those computers," discovered the value-free mayhem that later morphed into what someday may be known as "The Enron Nation." This book is an attempt to tell the stories of some of them—those who bounced back and those who simply splattered to the pavement.

What Happened?

Everyone, no matter how cynical or jaded they pretend to be, needs something to believe in. The Internet was this "something." It offered a way out. It was the American Dream

2.0: It didn't matter who you were or where you came from as long as you knew technology, were willing to break the rules, and were ready to work your ass off. Of course, the fact that you could also make a boatload of money and that the aging Flower Children didn't understand the Internet made it even more attractive.

Like many things that end badly, the Internet Gold Rush started off great. Every company that came down the pike seemed to be a winner, proving that Netscape wasn't just a fluke. First came Yahoo! and the other search engines. Then Amazon roared into our consciousness, bringing with it the tantalizing concept of electronic, friction-free commerce—a technological breakthrough that would supposedly liberate us from brick-and-mortar tyranny.

As time went on, stock prices rose higher and higher, and people became more and more insane. Playing the market became the national pastime, as hordes of day traders bought everything in sight with a "dot.com" at the end of its name and venture capitalists and investment bankers, eager to keep up with demand, unleashed scads of companies that under normal conditions would never have been funded, much less taken public.

If the CEOs fancied themselves rock stars, the money people behaved like a bunch of trend-obsessed TV executives, chasing popular taste and ending up with a series of hare-brained schemes: free ISPs, free computers, business to business, get-paid-to-surf, etc. It wasn't about perceived value anymore. Bullishness over a hot new industry became "irrational exuberance,"[1] which was Federal Reserve Chairman Alan Greenspan's polite way of saying, "total frickin' lunacy." The rationale, or, should we say, "irrationale," was that if

[1] Alan Greenspan, "The Challenge of Central Banking in a Democratic Society," remarks made at the Annual Dinner and Francis Boyer Lecture of The American Enterprise Institute for Public Policy Research, Washington, D.C., December 1996. Available: www.federalreserve.gov/boarddocs/speeches/1996/19961205.htm

Introduction

out-of-control Amazon was worth 400 bucks a share, then Joe's Breakfast Cereal E-porium was worth at least 100 bucks, provided, of course, that the investor was as high as a kite and intended to remain so for the foreseeable future.

With so much easy money at stake for bankers and brokers alike, due diligence went down the toilet, and the toilet went out the window. The $12 million in annual revenue that got Netscape out of the gate in 1995 became the underwriting standard until the Bubble's peak year of 1999, when theglobe.com pulled off a monster IPO with $12 million in annualized revenue, and Stamps.com went out even before it had launched a product, trading up at one point at a mind-boggling $98 a share. After that, it was truly "Anything Goes," with the funding of companies like JustBalls.com and Gazoontite.com defying all logic and common sense.

For those of us who had entered the industry several years earlier as an alternative to a boring job in the Old Economy, watching things turn into a virtual circus overrun by cyber-shysters, dot.con artists, and New Media groupies was more than a little disturbing. Sure, the pay was good, but the pace of Internet time was grueling and, after working at three different circuses in three years, we hoped that our latest job would be our ticket out of the big top, or, in some cases, the lion's cage. Some of us did, in fact, escape. Most, however, went on and on to the last possible moment, until it all came crashing down.

Why NetSlaves 2.0?

When we started the NetSlaves crusade against the evils of the Internet Industry back in 1998, people thought we were crazy. "Whaddaya mean this business stinks? I love my job. We're all going to be rich! Rich, I tell ya. Rich!"

Well, here we are over four years later, and people don't think we're so crazy anymore. No one has come out and admitted that we were prescient. But the shouting, all-caps e-mails, accusing us of being "JUST A BUNCH OF WHINERS," have stopped, and so have the funny stares whenever we complain about working the Web.

We'd like to tell you that we're above gloating. But we're not. And so to all you Ayn Rand libertarians, all you Henry Blodgett fans, all you CNBC worshipers, all you morons who thought the Dow was going to hit 50,000 and lost your kid's college money on theglobe, eToys, Pets.com, or some other worthless stock, a hale and hearty, "Ha, ha!"

This would be sweet revenge indeed, if it weren't for the fact that those responsible for the whole dot.bomb mess haven't exactly been dragged up and down Sand Hill Road (the legendary road where many West Coast VCs concocted many a get-rich-quick scheme) by the heels. Sure, some may end up in jail for playing strange games with IPO shares, but for the most part those wily CEOs, venture capitalists, Wall Street analysts, and investment bankers made it out of town with the loot, and we the people who built the Web are the ones who got screwed.

So much for the stock options. So much for the free pizza, the foosball tables, the Aeron chairs, the lunchtime massages. So much for "bring your dog to work." So much for "we're a family here." So much for "build it and they will come" and "get big fast" and "who cares about profitability?" So much for sock puppets, and selling toothpaste and furniture and BBQ sauce online. It's gone now, baby, and good riddance!

No, this isn't Schadenfreude. This isn't "we're devilishly delighted your dot.com slipped on a banana peel and broke its $100-million neck." Rather, it's about coming to terms with the madness of the past five years, so we can hopefully get beyond it. And not in some touchy-feely, group-hug kind of way, and certainly not in the hysterical ramblings of the crushed and the

Introduction

clueless. Forget the panicky screeds crying about the billions that were lost, and all the stupid companies that went out of business. You want to know what *really* happened? You look at what came before the insanity.

Flash back ten years, to the Bad Ol' Days of 1992. Papa Bush is in office. Nirvana is the biggest band in the land. And the U.S. economy is flatter than Iraq. As for the future, it looks a lot like a NAFTA Nightmare, with jobs being auctioned off to the lowest global bidder and workers left to fend for themselves in a world where loyalty and stability no longer exist.

Flash forward three years, to 1995. Sax man Bill Clinton is now the Entertainer-in-Chief, and it's as if the doom-laden, downsized, post–Cold War years never happened. The word on everyone's lips is "Internet," and the company on everyone's mind is Netscape. After literally coming out of nowhere and rocketing to the top of the IPO charts on a mere $12 million in revenue, Netscape captures the popular imagination with the intensity usually reserved for celebrities. In many ways, it's a lot like the sixties, only instead of a bunch of guys in a house in California saying, "Hey, man, let's form a rock and roll band, put out a record, and make a million dollars," you have a bunch of guys saying, "Hey, man, let's form an Internet company, go public, and make a billion dollars."

In retrospect, it's too easy to make fun of these young-sters, but at the time it seemed like a really good idea. After all, here was a generation that was supposedly so wretched that the media wouldn't even grace them with a name. They were "Generation X," and they were nothing but a bunch of slackers who, unlike the great Baby Boomers of the previous generation, would never amount to anything. They would either end up doped-out, with a bullet in their brain like Kurt Cobain, or else slaving away at McDonald's or some other dead-end job.

Did anyone stop to consider that there might be something terribly wrong in Wonderland? Not really. If you did, you just didn't "get it." You weren't hip to the strange buzzword language of "verticals," "vortals," "viral marketing," and so on. You were a non-believer, a Luddite, a fool. The media seemed particularly uninterested in reporting anything except the gee-whiz stories of twenty-three-year-old billionaires; the truth was, well, it just didn't sell newspapers.

Whereas the first NetSlaves book focused on the people working in the trenches of the New Economy, *NetSlaves 2.0* focuses on how a wide range of people dealt with the Wipeout. Those who've been most badly mauled are called "Lepers," those who've rejected technology are called "Neo-Luddites," and those who've attempted to reinvent themselves are called "Shape-Shifters."

What does the future hold for these people who joined the Net revolution with such high hopes but now find themselves burned out, thrown out, betrayed, or worse? What's next?

We don't honestly know. But the good news is that while many of them are unemployed and uncertain about the future, the stupid money is gone and so are all the poseurs who latched on to this industry just for the quick buck. We don't know about you, but to those of us who love technology and still believe in the Internet, no matter what the NASDAQ is doing, this is a big relief. Of course, mental serenity doesn't pay the bills, so that's why we're soldiering on with the NetSlaves site and the book you're holding in your hot little hands.

And here's something to keep in mind as you read on: The events, characters, and career crises in this book are based on real people and real companies. However, the names, both of people and dot.com companies, have been changed, and in certain cases events and characters have been "composited" and/or compressed. Still, most of this book is true. (We swear to God.) Ask the people we interviewed. Ask yourself. Ask history.

1

— Neo-Luddites: Take This Industry and Shove It! —

Neo-Luddites: Who Are They?

Neo-Luddites would rather shovel cow guts in a slaughterhouse than ever work in the technology industry again. Like their nineteenth-century British predecessors who rejected the mechanization of farming, Neo-Luddites regard the global efficiencies of the Internet as nothing short of the Devil's handiwork.

While this may sound like a radical view, straight out of the Unabomber's "Manifesto," Neo-Luddites do have a point. In many ways, the Internet hasn't improved our lives one bit. Not only has it turned the traditional 9:00 to 5:00 workday into a 24/7 digital hell of cell phones, pagers, and laptops constantly screaming for attention, it has also driven conspicuous consumption to new heights and brought us one step closer to an Orwellian future in which everything we do is recorded and later used against us, like when potential employers could order reports detailing a candidate's personal reading and buying habits online for the past five years. ("You visited AsianBabes.com three times in 1999. I'm sorry, but you're not the type of person who reflects the values of our firm.") These are valid concerns indeed, yet they are only half the story. The other, which Neo-Luddites have come to ignore, is that the Internet has enabled the free flow of public information like never before. Want to find out about human rights in China or check up on the environmental track record of corporations? It's all out there for the taking.

If the Internet isn't digitized evil incarnate, then why have Neo-Luddites rejected it? The reason is that many of them were so disillusioned by their experiences in the industry that the only way to reclaim their lives and their souls was to go to the other extreme and deny that they had ever been involved. Like reformed alcoholics who become violent if someone drinks a beer in their presence, Neo-Luddites are incapable of a rational middle ground because they are too caught up in their own coping strategy to see the bigger picture. You can call them "narcissists" or just plain "romantics," but Neo-Luddites are the type of people who either love a movie or think it's the worst piece of crap ever created. In a world where "lukewarm" is a valid rating for everything from presidential candidates to stocks, the passion of Neo-Luddites may seem quaint and perhaps a bit crazy, especially since they often end up detesting the very same person or object they used to adore.

As we see it, though, Neo-Luddites were very important to the development of the Internet. They were the Cyber-Evangelists who badgered you to get online and who believed in the democratic power of the Internet before anyone else did. That their intense optimism ultimately caused their intense disaffection is unfortunate, but Neo-Luddites are nothing without a cause, a principle that drives them forward.

Neo-Luddites attacked the Net with a religious and revolutionary fervor that didn't permit crass materialism. The Internet was about art, not commerce, and they were the artists, even if their contribution to the new medium amounted to nothing more than writing copy for Web sites or creating graphics for them.

Were Neo-Luddites fools to think that the Internet would never be big business? Of course. But they were no worse than those who succumbed to dot.com mania. At the same time, we do hope that Neo-Luddites aren't foolish enough to forever reject the Web just because a bunch of morons got greedy.

Are You a Neo-Luddite?

You might be a Neo-Luddite if . . .

- Your cell phone is now a twisted pile of plastic and your nineteen-inch monitor is the new dwelling of your pet goldfish.

- You think that *A Heartbreaking Work of Staggering Genius* is just that, and you find David Eggers's antics amusing. (We don't.)

- Suck.com used to be your favorite site before you trashed your computer. (See above.)

- You get misty when someone mentions long-gone Web zines like TotalNY, Word, Mondo 2000, and Cool Site of the Day.

- You remember cyber-soap operas ("The Spot," anyone?), "Mirsky's Worst of the Web," Packet, Flux, and AOL's Greenhouse Project to support homegrown sites.

- You begrudgingly acknowledge *The Onion*'s success because it should've been you who crossed over to the mainstream media and sold millions of books, T-shirts, and other doodads, not them.

- The Web didn't instantaneously die when the NASDAQ crashed in April 2000. As you see it, it suffered a slow and painful demise, beginning with the Netscape IPO in 1995, when people realized how much money could be made online and everything got commercial.

- These days, you fancy yourself a "working class intellectual" as you spout lines from Noam Chomsky to the proletarian masses. Instead of "building solidarity," however, you breed contempt as your fellow day laborers come to regard you as a College Boy pain in the ass who can't hammer in a nail straight. ("Hey, if you were as good with your hands as you are with your mouth, we'd have finished this roof five hours ago.")

Fun Facts About Neo-Luddites

What They Were Before the Web: Annoying, overly sensitive, overly intellectual, self-righteous academics. ("Do you realize that in order to maintain hegemony, the patriarchy must marginalize the discourse of the Other through . . .")

What They Were When They Got Net "Religion": Annoying, overly sensitive, overly intellectual, self-righteous Cyber-Crusaders, posing as Web writers, journalists, and copyeditors. ("The Web is a free space where metaselves engage in . . .")

What They Turned into After the Web: Annoying, overly sensitive, overly intellectual, self-righteous mistechnopthropes.

Current Employment Situation: Driving a taxi, waiting tables, etc. Unlike Lepers, who are doing these things out of necessity, they're doing them to make a point. (And what is that point? That you don't need technology, of course. Of course!)

The Dirty Little Secret of Neo-Luddites: They've watched *Office Space* about 300 times. (In light of their usual diet of French New Wave films, NPR, and Critical Theory, it's downright slumming.)

Post-Bubble Pain Rating (PBPR): 9.5, on our completely arbitrary ten-point scale. (Not that they notice or care.)

What Neo-Luddites Are Missing Out On: Now that all the stupid money is gone, a Web zine renaissance is taking place, with wacky, home brewed sites popping up on a daily basis, created by unemployed and underemployed techies who still believe in the subversive power of the Internet. (Our personal favorites: MrCranky.com, PitchForkMedia.com, and Fark.com.)

Neo-Luddites: The Story of Charles

It wasn't a war, but it certainly looked like one. Charles stood in amazement, watching the great machines shoot flames and lob bombs at each other. The smell of burning gasoline was everywhere, and much of the proud phalanx of treaded monsters already lay in smoldering ruins, torn to pieces in this orgy of screaming projectiles, or melted into rusty carcasses by robotic flamethrowers. Perhaps more insane than the spectacle itself was the fact that it was taking place in a parking lot in the middle of a densely populated area. Not surprisingly, it wasn't long before a squadron of police cars arrived, their sirens blaring.

"What the hell is going on here?" screamed an officer in riot gear.

"It's okay, man," one of the onlookers responded. "It's the Survival Research Laboratories."

"The what?! We had reports that the city was being invaded."

"It's all right, John," said another officer. "It's performance art."

"Jesus Christ, I'll never get used to this crazy town."

The police set about dispersing the crowd, then turned to the SRL folks, who had grown accustomed to such treatment. A paramilitary group of Texans with an aesthetic streak, the Survival Research Laboratories had been producing events around the world since the late seventies and had become heroes in the San Francisco tech community, and quasi-criminals in the eyes of the police.

"You guys are a bunch of freaks, you know that?" barked an angry sergeant.

"Yeah, whatever, dude," said one of the SRL organizers.

The ire of the police aside, the SRL was strangely never more relevant to what was happening in the Bay area. In a metropolis besieged by the Internet Boom, where you couldn't walk two feet without seeing a dot.com ad or ride in a taxi that wasn't emblazoned with a goddamn logo, watching homemade robots blown to bits made every fiber of Charles's being tingle with joy.

"Fight Club" for geeks? A metaphor for the tech industry as a whole? A communal cleansing of class rage? Charles's mind raced to find the perfect reference to explain what was happening. But no matter of learned associations could capture the primitive joy that he and the rest of the crowd were experiencing, nor could even the most prescient of them have predicted that it was a vision of what was to come. Technology, hailed for the past five years as the Great Redeemer, would soon morph into a scorching, fire-breathing Destroyer that would make the controlled Apocalypse they'd witnessed look like child's play.

On a certain level, Charles realized that big changes were afoot. His life in the Promised Land of San Francisco had grown so intolerable that something had to give. He had lost his girlfriend, he hated his job, and most significantly, he had stopped believing in the Internet. "What next?" he asked himself over and over again. "What next?" The problem for Charles was not what he didn't want to do, but what he considered worthy of doing. He had given himself passionately to both his career and his relationship and, at twenty-six, he felt a profound emptiness in the center of his chest where there should've been a purpose.

Tonight was no exception. Driving back from Oakland, where the event had taken place, he felt especially low, and the only thing that kept him from slamming into a girder on the bridge was the thought that his story about the SRL would be his epitaph and his final statement on the "Digital Revolution."

Long Live the King

Charles had arrived in the shining City by the Bay only three years earlier. Like many young people fresh out of school at the time, he was full of dreams about the Internet. For him, though, it wasn't the chance at a moon shot IPO or even an underlying love of technology that was the attraction. The only child of two left-leaning sixties parents, Charles had much loftier ambitions. As he saw it, everyone his age who was rushing into the industry for the big bucks was missing the point. It wasn't about stock options or a fat salary; it was about changing the world and using the universal publishing platform of the Web in order to do it.

To folks already working in the business, Charles's ideals would've seemed as laughably archaic in the Internet year of 1998 as the nose ring and the phrase, "If you build it, they will come." Of course, anyone crazy enough to try to convince Charles otherwise would've gotten an argument. He would've proudly pointed to the high level of editorial being produced by sites like FEED, Suck, Word, Salon, Slate, and his future employer, *Wired*.

While he wouldn't have been wrong in stating that these were fine examples of digital publishing, he wouldn't have been right either. For one thing, it wasn't ad-supported culture zines that were getting the funding and the media attention anymore; it was e-tailers like Amazon.com, CDNow.com, and Pets.com—companies that had taken the Web to the next level by showing that you could make money selling things online, at least theoretically. For another thing, none of the content-driven sites were making enough money from advertising to support themselves. As a result, many tried boosting revenue with subscriptions, with almost universally disastrous results. TheStreet.com tried them and failed, and so did Slate, and later on Salon.com and Inside.com. Even the Wall Street Journal Online, which had long been the poster boy for electronic subscriptions, ended up having to lay off a considerable number of staff members during the ad crunch of 2001.

Charles had a valid excuse for being so clueless about such things. He was an academic from Kent State whose introspective personality and literary pursuits had so far shielded him from reality. He would find out soon enough how wrong he was. In the meantime, however, he remained blissfully unaware that it wasn't 1995 anymore, a time when even big media companies believed that "Content Is King." First among them was News Corp's incredibly short-lived iGuide, which started as a general catalog of all things Internet as well as an online home for several of the company's media brands that rose to momentary glory, before collapsing into a placeholder for TVGuide.com. Then, of course, who could forget Time Warner's Pathfinder, a gorgeous mosaic of more than eighty trusted brands that, after years of gross unprofitability and internecine rivalries, forced the conglomerate to choose between oblivion and the sheltering arms of AOL. Even AOL, which up until then had been nothing more than a proprietary, obscenity-driven chat client, had made bold moves into Content by launching its "Greenhouse" project, an effort that eventually spawned such sites as Motley Fool, Hecklers Online, and Astronet. In the dream days of 1995, things were so silly that Microsoft—a company whose experience with content had been limited to deceptive press releases and confusing menu bars—embarrassed its Netscape-hating, anti-competitive self with Mungo Park, an online travel guide that tracked celebrities as they traipsed around the globe, showering the natives with product placements. The net effect of these massive corporate initiatives was to create a generation of people like Charles—Comp Lit and other Liberal Arts majors whose word skills were suddenly in demand—to write, edit, code, produce, and otherwise populate the Web with fresh, original, and interactive material. While their masters viewed online publishing as merely a new distribution channel, to Charles and his fellow slaves its advent represented nothing less than a new literary form.

A Portrait of the Artist as a Young Futurist

It was back in Kent, at the urging of the manager of the used bookstore where he worked, that he began reading *Wired* cover-to-cover and attacking such texts as John Perry Barlow's "Intellectual Property in the Age of Cyberspace," Douglas Rushkoff's *Playing the Future*, and most importantly, everything written by and about Marshall McLuhan. What shocked his manager was not how quickly he absorbed such radical futurism, but that he took to it at all. Until then, his tastes had been eclectically divided between Russian novelists, the radical theater of Alfred Jarry, Michel Foucault, John Barth, and, of course, the Beats. When asked by his fellow classmates why he had abandoned literature, his response was, "Don't you see? This is the New Literature! A literature for the people! Everywhere. No more photocopying your zine at your job and hoping that fifteen people in your neighborhood read it. The Web gives you a worldwide audience instantly. All you need is a computer!"

Dismissed by many of his classmates as someone who had latched onto the latest Marxist fad, Charles nonetheless continued his quest. He taught himself basic HTML, devoured the entire history of computing from the abacus to the ENIAC, and cranked out abstruse essays on nano-consciousness. Borrowing his manager's WELL account, he posted many of these tracts to the Futurist Conference and soon found himself embroiled in intense debates with San Francisco's digital elite. One of the participants turned out to be a senior editor at *Wired*, who was so impressed by his work that she invited him out to San Francisco "to discuss possibilities."

"What should I do?" he confusedly asked his manager

"Uh, that's a no-brainer, Charles," she answered. "You go."

"But where will I get the money for a plane ticket? And where will I live?"

"I'll lend you the money for a plane ticket. And you can live with my niece, who's a freelance designer."

"Are you sure?"

"You'll be fine. Vivian knows the city very well—maybe too well."

"Oh, really?" he said, his curiosity piqued.

About a Girl

A few days later Charles was in San Francisco's SOMA district—a hardscrabble industrial area whose many warehouse spaces already housed hundreds of fledgling dot.coms. He paced back and forth in front of what looked like an old candy store. "This can't be the right address," he mumbled to himself. "This can't be it." Several hours passed and Charles was about to go looking for a cheap hotel for the night when a taxi stopped in front of the store and out stepped a young woman with bright pink hair and a backpack.

"Are you Vivian?" Charles said somewhat desperately.

"I am," she responded tentatively, then continued. "You must be the Kent guy. Hey, I'm soooo sorry I kept you waiting. I'm just getting back from Burning Man."

"It's cool. Better late than never! Let me get that for you." Reaching out to take her backpack, he got a whiff of the strongest body odor he had ever smelled. As much as he tried to remain open-minded about such things, he couldn't help gagging.

"I know, I'm ripe," she giggled. "Such is the price of Primitivism."

Charles laughed. "Let's go inside. You must be tired. I certainly am."

Vivian's place turned out to be exactly what it seemed to be—an old candy store that had been converted into an apartment. In true hipster fashion, there were posters and paintings lining the walls, a makeshift kitchen in the corner, innumerable candles, piles of books everywhere, and a rear storage area that served as Vivian's bedroom.

Taking a seat on the oversized couch in the middle of the room, Charles was in his glory. "This is so bohemian," he said to himself. "This is so fuckin' bohemian."

"Would you like some tea?" Vivian asked, interrupting his reverie.

"Yes, please," he replied with a smile, then thought, "I'm sleeping on the couch tonight, but something tells me I won't be for too long."

Charles was right and wrong again. He ended up in Vivian's bed—actually her mattress on the floor—that very night after they had sat around smoking pot and talking until 3:00 A.M. Even though Vivian still hadn't bothered to take a shower, it was very sweet nonetheless, especially since Charles hadn't been with a woman in quite some time. Vivian, who seemed to possess a worldliness beyond her young years, picked up on this immediately and didn't hesitate to tease him.

"Gee, was I THAT awful?"

"No, dearie. Just a little rusty. But don't worry. I'm more than happy to help you practice."

In the days that followed, Vivian did much more than work on Charles's lovemaking skills. She bought him new clothes because she said he looked "like a grunge boy from up North," she encouraged him to shave off all his hair, and she even convinced him to get a tattoo of the yin and yang on his arm, which she also paid for.

"You look great," she said, as if she were marveling at a Web site she had redesigned. "You look like you've been a San Franciscan for years."

"Really?"

"For sure. They're not just going to hire you; they're going to give you a really good job. You'll see."

As Charles would later learn to his dismay, Vivian was always right with her predictions. That Monday, the editorial team interviewed Charles for about fifteen minutes before offering him the Culture Editor slot for Wired News, the magazine's Web site and favorite destination for futurists and geeks alike. Charles was beside himself with joy, but tried to seem disinterested for fear of blowing his new-found cool.

He called Vivian that afternoon to tell her the good news and, when he found out that she had talked him up big-time to the managing editor, he almost uttered those three little words.

The Best Minds of My Generation

For the next several months, Charles plunged into the Culture beat with a vengeance, pouring his heart and soul into his work. While the rest of the staff was either glued to the TV screens that surrounded the newsroom or surfing the Web for leads, Charles seemed to pluck story ideas out of the "noosphere"—that unseen part of the universe where the future became the present. Under his editorship, the Culture area became ground zero for such hot issues as MP3 zealots seeking to take on Warner Bros., the latest guerrilla media antics of AdBusters, and Open Source crusaders who were building computers for inner-city youths. To Charles's surprise and delight, many stories he suggested, wrote, or edited quickly found their way from the Web to the hallowed pages of *Wired* magazine, an achievement that made him one of the rising stars at *Wired*, destined for a bright career as a writer and pundit.

On the personal front, things also couldn't have been better. He was head-over-heels in love with Vivian and, having money for the first time in his life, he leased a brand-new Jetta, whisked her away on day trips up and down the coast, went out for expensive meals at the city's best restaurants, bought stacks of books, hung out with like-minded friends from *Wired*, and assembled an impressive CD collection that spanned everything from the latest Jungle mixes and Gamelan chants to Thelonius Monk and Stereolab.

Charles, for lack of a better word, was happy.

Cloud

Charles's idyll in digital paradise ended when the media empire that Louis Rossetto and Jane Metcalf had built stumbled. Although its glossy multicolored magazine was making money, its other media properties, including its publishing arm HardWired and its sprawling online assets, were bleeding red ink. For a time, there had been talk of an IPO to raise money, but with the market still reeling from the Asia Crisis and the company's finances in such dire straits, the only option seemed to be an acquisition.

After months of rumors and speculation, Charles and his fellow staffers were gathered into a conference room in May 1998 to be given the "good" news. The empire would be divided in half, with *Wired* magazine going to East Coast publishing giant Condé Nast and the sites remaining independent until a suitable buyer was found. In essence, there would be two Wireds, operating as separate companies, yet "cross-marketing" and "building synergy" wherever possible.

Charles left the conference room in a stupor. What did it really mean? How much of it was true and how much was just PR? Would he still be able to run pieces in the magazine?

Charles's worst nightmares came true a few weeks later when he tried to buy lunch in the Condé Nast cafeteria up the block from Wired News's office.

"I'm sorry, sir, I'm not allowed to serve you," said one of the counter people.

"What are you talking about? I'm from Wired."

"Yes, but not *Wired* magazine."

"This is bullshit," he screamed, slamming down his tray.

Charles stomped off vowing to write the most scathing e-mail he could muster, but when he returned to his desk to begin the missive, he noticed a memo in his inbox, stating that "Wired News employees are no longer welcome in the Condé Nast dining area."

"Well, so much for synergy," Charles snapped to his cube mate.

"It might be time to get the hell out of here."

Charles didn't respond. He didn't even want to consider such an option. He had come so far, he had done so well, there had to be something he could do.

There wasn't. All his e-mails went unanswered; all his calls went unreturned. After a while he didn't even know who to contact over there anymore because the people he knew had left or were planning on leaving. Condé Nast's true intentions had finally become apparent. They didn't want anything to do with Wired Digital. As far as they were concerned, Wired Digital was East Germany and they were West Germany, and never the twain shall meet. It was the magazine's circulation numbers and fat ad accounts that mattered, not a bunch of money-losing Web sites. Technology and comarketing be damned!

Charles sighed at this revelation and watched in horror as the magazine got fatter and fatter and came to resemble Condé Nast's other properties such as *GQ, Vogue, Self,* and *Vanity Fair*—a "lifestyle magazine" for the upscale digital set.

Charles found the whole thing disgusting. But, as his colleagues started heading for the hills, Charles soldiered on, publishing the same thought-provoking stories that had become his trademark, in the hope of single-handedly preserving *Wired*'s good name.

And then things really got bad.

Invasion of the Lycosians

In October 1998, *Wired*'s digital assets were purchased by none other than Lycos, the Waltham, Massachusetts-based search engine company. When Charles heard the news, he knew that his days of plucking ideas out of the noosphere were over. Although he fully expected to be fired, he soon found himself suffering an even worse fate.

Since Lycos was a publicly-traded company, and one that knew nothing about publishing, editorial took a backseat to utility and performance: the old numbers game. The question no longer was, "Did Charles write an interesting story?" but rather, "Did Charles's story get a lot of traffic?"

To be fair, nothing that any of the Wired News staff wrote made any difference because even their most popular and insightful reporting never measured up to the millions of eyeballs that Lycos and Wired's HotBot search engine attracted on a daily basis. They had all become second-class citizens in a company where they used to be the darlings.

A mass exodus ensued. Many left in a huff without even thinking about where they would go next. Most assumed that their genius would be better appreciated elsewhere on the Web. Salon, a San Francisco–based site with delusions of content grandeur, was inundated with résumés submitted by hopeful applicants. A few were lucky enough to make the transition; the remainder wound up in low-level editorial jobs, stoking the flames of e-commerce with promotional copy and product descriptions.

Scared of winding up like his former colleagues, Charles stayed put. But soon he began to feel as if the ground was slipping away under his own feet. First, he got pushed off the Culture beat. Then, as a regular reporter expected to cover hard business news, his contempt for the industry's IPO mania was so apparent in his writing that his editor offered him a choice: Become the WebMonkey editor or hit the bricks. Charles chose the former and soon found himself poring over WebMonkey's pragmatically-oriented set of tutorials that ran the gamut from basic HTML to advanced PHP programming. He despised the work, yet it beat writing puff pieces about CEOs and companies that sold nose spray online.

To keep himself sane, Charles was able to slip an occasional Culture piece under the wire, thus bringing back the old spirit of *Wired* for a flashing moment. But these moments grew few and far between, and one day while he was editing a substandard piece that would never have run in the old days, but now passed for brilliant work from the gang of unknown stringers who had been brought in to replace what was once a tight editorial team, Charles finally admitted to himself that he was miserable and needed to do something.

Home Is Where the Hate Is

Charles turned to Vivian for support. But she was even worse off than he was.

"I used to do good work, interesting work. These days it's all e-commerce. Which is a nice way of saying, 'retail.' Well, I didn't go to art school to draw pictures of shoes. I didn't perfect my skills to become a glorified sales girl. If I'd wanted to work at a mall, I would've stayed back in Michigan."

Charles stared at the floor, not knowing what to say.

"Don't pout. I hate it when you pout. I'm trying to talk, don't you want to talk?"

Charles exploded.

"If you want to talk, then talk TO me, not AT me. This isn't a conversation, this is a rant. If you . . . "

"GET OUT! JUST GET OUT!" Vivian screamed.

Charles grabbed his coat and slammed the door behind him. He walked the streets for hours, now with yet another problem on his mind. His relationship with Vivian hadn't been good for months. They rarely talked, they didn't go out anymore, they were always busy with some crazy deadline and never seemed to have time for each other. He still loved her, and she still seemed to love him, but it was like she was ashamed of what he did for a living. She fell for an artist, not some glorified copy boy, and whenever he tried to convince her otherwise she wouldn't listen. She was too busy projecting her own frustrated aspirations onto him. At least, this was his hypothesis.

He returned to the apartment around 2:00 A.M. There was no trace of Vivian except for a note written in purple ink that had been left on his pillow.

> The Dream is gone and so am I. I guess I should've told you this sooner and to your face, but I'm going to be in Belize for the next six months or so. This industry's gonna crash, and I don't want to be anywhere near it

when it does. Please forgive me for the way I've treated you lately. I love you, you're a great guy, even though you have to get up off your self-pitying ass and do something about your life.

Charles cried for the rest of the night and showed up the next day for work in a fog. He remained in this state for several months and, in moments of weakness, wrote Vivian six-page e-mails, none of which she responded to. When he had just about given up on her, she called him out of the blue one afternoon.

"It's over, you know."

"I kinda figured that."

"Good."

"When are you coming back?"

"I don't know. But when I do, you're going to have to move out."

"I understand. How can I reach you?"

"The best thing to do is leave a message with my parents. There are no Net connections down here and it wouldn't matter even if there were. I traded in my laptop for dope."

He was about to ask her if she had been smoking too much when the line went dead. Charles weakly put down the receiver and was on the verge of tears as his editor strode up to his cube.

"Are you okay?"

"Yes, fine. What's up?"

"I need someone to cover the SRL event tonight."

"Sure, I'm on it."

Killing Myself to Live

Any joy Charles may have felt that afternoon was completely gone when he got home that evening, and he was more resolved than ever to finish the story, and his life. Firing up his laptop, he stared at the blank screen for a moment, then went to work. Although it had been a while since he had written something substantive, the words came gushing out in a white-hot fury interrupted only by the repeated, unwanted appearances of Mr. Paperclip asking, "Are You Writing a Letter?" When he was done, only an hour and a half later, there was a ten-page rant looking back at him, the very sort of impassioned tirade that had gotten him the job at *Wired* in the first place. He was almost certain they wouldn't run it, but he e-mailed it off anyway and proceeded to his next task: suicide.

How would he do it? Jumping off Coit Tower or the Golden Gate Bridge was too obvious, guns were messy, he really wasn't into hard drugs; so what would it be then? He finally decided that he would gas himself. Walking slowly over to the stove, he turned all the burners on full-blast and went to bed.

Around 11:00 A.M. the following morning, the phone rang and, miraculously, he woke up and answered it.

"Uh, hello?"

"I can't use this piece."

Charles struggled for a moment, his head splitting, before recognizing his editor's voice.

"I don't care if you use it or not. I quit."

"Is there any way I can get you to reconsider?"

"I'm afraid not."

"Lycos just announced that it's being acquired by Terra Networks. What should we do about your stock options?"

"Do I sound like someone who cares about stock options?!"

Charles slammed down the phone without waiting for an answer. Stumbling out of bed, he made his way to the kitchen through an ever-increasingly potent gas cloud. Holding a dish towel over his nose and mouth, he shut off all the burners and went to the windows, but found them all already open. Realizing now why he wasn't dead, he began laughing out loud, and was strangely relieved.

He spent the day walking around San Francisco, a thing he'd been too busy to do for a long time. Starting with Fisherman's Wharf and the Embarcadero, he eventually hiked up Columbus Avenue to City Lights Books. Faces from a bygone era stared at him judgmentally from dusty frames on the walls: Kerouac, Ferlinghetti, Cassady, Corso, Snyder. With their gray, yellow lips, they all seemed to be calling him "a sellout." Depressed, he fled the towering stacks and, after climbing more hills, found himself in MacArthur Park. He sat on a bench as the shadows lengthened. People passed him as if in a dream. Some talked on cell phones, others were in small groups, beaming business cards among Palm Pilots, doing their best to ignore the occasional homeless person begging for change. Taking a fresh legal pad from the bag he was carrying, Charles attempted to record his impressions. At first, his mind was as blank as the page in front of him. But suddenly it all made sense: the wasted money, the wasted youth, the dreams of all the people he had known—and the people he hadn't known—the rushing, insane desire to ascend to the loftiest heights of the American Dream and everything it promised the human spirit. His hand trembling, Charles was about to set forth the first word of his grand opus—the one that only came into a writer's mind once in a lifetime. Melville had his "Call me Ishmael," Joyce had his "Stately, plump Buck Mulligan," and Dickens had his "It was the best of times, it was the worst of times." And now it was Charles's turn. But just as the tip of his pen was about to touch the paper, a Yahoo!-emblazoned cab, blaring its horn, came screeching out of a side street, and jumped the curb.

To avoid losing his legs, Charles hopped over the side of the bench. The legal pad went under the cab's wheels, and the pen went skittering across the pavement. By the time Charles had righted himself, both the cab and the first words of his universe-shaking masterpiece were gone.

When he got home that night, he gathered all his gadgets into one big pile and started smashing them to pieces with a hammer. They included his laptop, his cell phone, his Palm Pilot, even his answering machine. If he'd had a TV and a VCR, they would've been gone too. As it was, he had more than enough technology to take his aggressions out on, and by the time he had finished his one-man assault, he stood knee-deep in electronic debris.

Charles probably didn't realize it, but he had become a Neo-Luddite.

No Apologies

Charles spent most of 2000 passing from one low-paying, no-tech job to another and writing wherever and whenever he could. Although he wasn't aware of it, he was hardly alone. Thousands of other once-aspiring Web scribes found themselves in equally desperate straits, forced by necessity into office temping or other dead-end administrative jobs. Others who shared Charles's extreme mindset were leaving San Francisco in droves, signing up for the Peace Corps, returning to school, or losing themselves in extended vacations overseas. By the beginning of 2001, so many people were moving out of the Bay Area that real estate values plummeted, and local U-Haul branches had one-month waiting lists for anything with wheels. Charles still kept in touch with several of his *Wired* friends, all of whom were shocked to learn that he had joined the ranks of the digitally disaffected, and was now stooping to such menial jobs as dogwalking, dishwashing, roofing, and waiting tables. While many of his former colleagues weren't doing much better than he was, they hadn't stopped trying to make it in the industry.

Charles looked down on his friends for what he saw as a lack of courage—a refusal to face the fact that the industry was fucked. The stupid money was gone, but the streets were still filled with believers hoping that the New Economy would quickly bounce back. As Charles saw it, these deluded people had learned nothing from their experience and would rather go on humiliating themselves for a paycheck and a warm keyboard to attempt something they truly loved. He took great joy in seeing the companies he had ridiculed go up in flames in the great dot.bomb conflagration, which confirmed everything he thought about the industry and had seen foreshadowed in the flames of the Survival Research Laboratories event. Meanwhile, his Schadenfreude and contempt didn't prevent him from borrowing money whenever he was running low, which was most of the time. Being writers themselves, his friends lent him whatever they could spare and encouraged his literary efforts, while gently prodding him to accept slightly more "prestigious" work.

"Look, I heard about a Web copywriter job at Yoo-hoo, you interested?"

"Yahoo!?"

"No, Yoo-hoo, the bottled chocolate milk people."

"Forget it, Tom. Thanks, but no thanks. On the other hand, if you told me that the *New Yorker* or, better yet, *McSweeney's* was looking for a new poetry editor, then . . . "

Such were Charles's aspirations. And he worked toward them with the same determination he had once applied to building up Wired's Culture channel. Not a day went by that he wasn't found in the post office, mailing something out to a major magazine or literary journal, or composing a witty query letter that he hoped would catch some editor's eye. For the most part, he never heard back from anyone, and his writing was returned months later in pristine condition, as if no one had ever read it. Occasionally, though, there were a few encouraging words scribbled at the bottom of a page such as "nice phrase, keep at it," but thus far, nobody had liked anything he was doing enough to actually publish it.

Charles dismissed these rejections the way most struggling writers do: "They just don't get me," "I'm too avant-garde," "Ginsberg went through the same thing." Even his biggest supporters had to restrain themselves from laughing at these rationalizations, yet they couldn't deny his dedication. He was working at an outstanding rate, churning out short story after short story and poem after poem, like it was being dictated from some invisible source. According to his calculations, he had written enough for three books since leaving Wired. It was no small achievement and one he hoped would finally be worth the price.

Into the Street

As always, Charles was up late one night pounding away on his manual typewriter with the local jazz station going in the background, when he heard a key turning in the lock. Before he knew what was happening, a bronze-skinned woman was standing in the middle of the living room with her hands on her hips. At first he didn't know who it was, but when she began shaking her finger at him, he realized that he had come face-to-face with the Inevitable.

"Vivian, wow, it's you!" he said, trying to make small talk. "Your hair's so different, so long, so brown . . . "

"Cut the crap, Charles, you know what this means."

He was about to ask her to at least let him stay the night when a large, shirtless Indian dude, straight out of a Jim Morrison desert hallucination, appeared in the doorway and flexed his overdeveloped pecs at him.

"Okay, I can take a hint," Charles said sheepishly, and proceeded to jam some clothes into a bag along with a stack of writing and a few books. Vivian, meanwhile, didn't waste the opportunity to verbally grind him into the ground one last time.

"You really look like shit. You're all skin and bones. Are you starving to death or something?"

Charles didn't respond and, when he was finished packing, he walked out the door with the bag in one hand and his typewriter in the other. It wasn't until he reached the sidewalk that he screamed, "Have fun with your shaman, bitch," before heading off to the backseat of his Jetta to sleep.

Epilogue

In 2001, Charles really hit rock bottom. He bounced from one friend's couch to another and eventually had to get rid of his car and his prized CD collection just so he could afford rice and beans. By the summer, he had grown so tired of picking up dog poop and washing dishes and not having a place to call his own that he decided that he needed to get a real job, or else wind up in a homeless shelter or, even worse, back in Kent. For a brief moment, he flirted with the idea of writing for another tech magazine, but after poring over *The Industry Standard, Red Herring,* and *Business 2.0* at Borders one afternoon while sipping a latte, he felt himself becoming physically ill. Not only did his brand of cultural analysis have no place in such business-obsessed publications, it seemed completely irrelevant to an industry interested only in Wireless Middleware, Broadband Penetration, and what companies were doing to survive the slowdown. Even if Charles had gone down this path, he would have found himself unemployed once again because all of these publications—and many more like them—that had sprouted like weeds during dot.com mania had gone out of business or shrunk to insignificance by September 2001.

With almost no money left in his bank account, Charles's luck finally changed when he landed a job teaching composition at a local community college. It may not have been as glamorous as working for *Wired,* but it allowed him to move into a tiny studio apartment and begin work on his latest project: a mock-heroic verse epic about his online experiences, loosely based on Milton's *Paradise Lost.*

— **Panhandlers: Will Code for Food** —

Panhandlers: Who Are They?

After the dot.com bubble burst, common wisdom had it that the only people making money were the liquidators and Pud from FuckedCompany.com. This is patently untrue. There was also an army of Independent Consultants who were making a comfortable living for themselves building mid-sized Web sites for clients who were either sick of getting ripped off by the big agencies or simply weren't worth their time.

Sure, a $30,000 project here and a $20,000 site there is chump change if you're Razorfish or Organic. But if you're one guy in a room with no overhead, except rent and a monthly ISP bill, it's quite a windfall, especially since, after years of breaking your ass for worthless stock options, working on your own terms is a gift from heaven.

We know this joy first-hand because it's exactly how we survived the immediate aftermath of the Bust. The irony, of course, is that the very thing that was killing off companies left and right made 2000 the most prosperous year of our lives: We cranked out site after site and ended up feeling like quite the entrepreneurs. Unfortunately for us and many other people like us, though, our good luck didn't last. 2001 turned out to be the mirror opposite of 2000. The clients who came to us so easily through friends and contacts simply stopped coming, and the few who did didn't want to pay for anything. The same work that would've netted twenty grand just a few months before was now going for rock-bottom prices. In fact, clients no longer asked us how much something would cost, they simply said, "I've got $1,000. What can you do for me?"

It was an awful situation. And for those of us who had gotten used to waking up at noon and being paid handsomely for a few days' work, it was a painful revelation. Make fun of us if you want. But how would you feel if you opened your own business with the idea of providing good service at competitive prices, only to have the market turn right around on you and declare that you are no longer an Internet Consultant, you're simply a Panhandler, a NetSlave of a different kind?

No, we're not kidding here. While Pud and the people selling off Aeron chairs got all the press, there were thousands of us below the radar who had the rugs pulled out from beneath us yet again. As if the embarrassment of crapping out of the Web industry hadn't been bad enough, we now had to contend with a new, more personal failure. And this time we couldn't blame company management or a stupid IPO; it was all on our shoulders. If we lost a client or got stiffed, or didn't "get it in writing" or keep proper tax records, we were the ones to blame.

How did we screw up? Well, we were so glad to escape our bonds that we didn't stop to look at what we were getting ourselves into. Our dot.com cluelessness begat a post-dot.com cluelessness. We were haughty enough to think that our series of lucrative, but difficult, jobs in the Net biz was all that we needed to make it as small business owners. As a result, we behaved much like the people we had formerly criticized: We blew our money on bullshit, didn't sock away enough to pay our taxes, and thought that our little scam, our comfortable little cottage industry, was invulnerable to the whims of the market.

And when the bottom of the bottom fell out, and we scrambled for any project we could get, we were perhaps finally getting the lesson we had so far managed to avoid learning. Running a business, even if you're one person in your underwear, is not for the faint of heart, and all those notions about the flexibility and greatness of self-employment are utter bullshit. It takes super-human persistence and a great deal of luck to make it on your own if, of course, you don't mind living as a Panhandler.

Are You a Panhandler?

You might be a Panhandler if...

- Your dream of being an Independent Consultant has left you broke and crazy (as opposed to well-paid and crazy, which is what you were when you had a job).

- You've gone from making boatloads of money building Web sites to scrambling for change from sleazy strangers you'd be afraid to work with under normal circumstances.

- You tend to say things like, "Have you considered that having a Web site for your (INSERT—pizzeria, cleaners, hot dog stand, etc.—HERE) might really improve business?"

- Although your bank account is rapidly approaching zero, you still think there's a project out there that will magically appear to save you from the fix you're in.

- When a client calls and asks how business is, you always tell them, "Great! Couldn't be better!" even though you haven't had a real project in months.

- You're too ashamed to tell your family how poorly you're doing. As far as they're concerned, you're the next Bill Gates. (Little do they know you don't have health insurance or a retirement plan, and you're living on peanut butter and jelly sandwiches.)

- You've fantasized about bankruptcy on more than one occasion. The only problem is that, now that the laws have changed, you don't even have that option open to you.

- As badly off as you are, you don't regret going out on your own for one second. For you, it was the only reasonable alternative to working sixteen hours a day as a stock option slave. (You are a noble ass, to the very end.)

Fun Facts about Panhandlers

How They View Themselves: The Clint Eastwood of Code; the tough, nameless, mysterious lone gun who comes out of the desert and rescues the town from the Bad Guys (aka, the in-house development clowns who have turned things into "a circus.")

How Other People View Them: Overpriced, interloping know-it-alls who should've stayed under the rock they crawled out from.

Psychological Profile: Fiercely independent types who despise taking orders from anyone, particularly middle-management generalists who attempt to control every step of development, even though they have no idea what they're doing.

The Animal They Are Most Like: If they were animals, they'd be wolves because they are predators who are either wandering around on the verge of starvation, or flat on their backs in a daze after sating themselves on an entire yak (the client).

What They Did During the Web Boom: Many freelanced their way through the entire period. The rest were digital prima donnas, notorious for throwing fits if someone changed the spec or pulled deadlines out of their ass without checking with them first. (The IPOs and the prospect of getting rich were beside the point. If anything, they viewed the people who were obsessed with these things as "poseurs.")

What They Did After the Boom: Set up their own shops and lived well until the work dried up.

Current Employment Situation: Most have begrudgingly returned to regular jobs; a few are still clinging to their dream of self-employment. ("Yeah, I know I haven't gotten paid in three months and my last three clients screwed me, but I love being my own boss.")

Panhandlers: The Story of Caleb

Caleb sat across from the accountant, dimly aware that the man had just said something, but unable to grasp its full meaning.

His mind was far away on some kind of animated treadmill accompanied by a hip-hop loop where cartoonlike creatures popped up, break-danced, and leaped out of the frame. But then the projector in his mental screening room unexpectedly ran out of film.

"Excuse me?" Caleb said. "I must have missed what you last said."

"I said that in the year 2000 you made $144,568 and you haven't paid a dime in federal, state, or local taxes," the accountant said, in as sympathetic a tone as was possible.

Now the animation came back again and Caleb could see animated Revenue Agents perched alongside the unwinding film, dancing to the Devil's violin, poised to pounce on him with rope and long hooks. Caleb was running—running on the film—but with each frame they'd gain ground, until the screen collapsed in blackness.

"Mr. Roundtree, are you all right? I know it's quite a blow."

"How much do I owe the tax people?" asked Caleb, banishing the animation from his mind.

"That depends on how much you can prove you spent," the accountant said, consulting his tax rate schedule. "You're single, correct?"

"Yes."

"No kids or other dependents?"

"Not unless you count my turtles."

"No turtles," the accountant said, punching numbers into a calculator. "$14,381.50, plus 31 percent of $61,018 . . . equals . . . $18,915.58 equals, uh, a grand total of $33,297.08."

"Jesus."

"And that's just Federal. You have to pay California State and local taxes as well."

"Oh my God. What do I do?"

"You have to pay them but, again, I don't want to paint too bleak a picture here. I'm pretty sure I can bring your tax liability down significantly, but I also need your help. The first thing I'm going to do is file an extension for you," said the accountant. "I'd also like you to execute this form, giving me power of attorney so that I can try to work out a deal with the IRS."

"Okay."

"I also need you to do some homework. Basically, dig up every receipt that can show any possible business-related expense that you've had in the last year. Doing so can cut your tax liability down substantially. I need receipts for phone calls, gas, car rental fees, W9's for the freelancers you paid, ISP expenses, education-related expenses, even rent, if you did much of your freelance Web work at home. Did you keep track of these expenses?"

"Yes and no."

"Yes meaning you had these expenses but didn't keep good records?"

"Yeah. I've been too busy. You know how it is."

"Dig them up. And do it sooner rather than later. I can keep the IRS at bay for a while, but their patience is not unlimited. Sooner or later, you'll have to pay the piper. And the longer you wait, the higher the additional charges are. You know, statutory penalties, accrued interest compounded daily, and other miscellaneous charges."

Caleb signed the form and walked outside into the 105-degree heat of West L.A. After adjusting his eyes to the glare, he staggered to his Acura, opened the door, and sat down at the wheel, his brain baking in the overheated air. It was 2:30—enough time to get to the bank to put in a check that would cover the $300 he'd had to pay the accountant to put him on retainer. But even the check in his pocket—$1,100 he'd been paid to put up a client's Web site—wouldn't make more than a tiny dent in the huge debt he now owed the government.

He cursed the World Wide Web, the year 2000, and his own feckless stupidity. How, he asked himself, could the most talented Web designer in L.A. have fallen into a ditch that he'd probably never dig himself out of without robbing a bank?

Tinkerer

From an early age, Caleb had drawn better than most kids in his class, and his exquisitely detailed action cartoons often adorned the corridors of the private elementary school he attended in Pacific Palisades. He discovered computers in seventh grade and soon began merging his artistic talent with the personal computer after he fell in love with an early Atari model that the school had purchased for its Art Department. By the time he was in eighth grade, Caleb had persuaded his mother to get him a brand-spanking-new Commodore 64 with tape drive. Soon he'd moved on to the Amiga. It was the first home computer with a serious operating system, million-color graphics, and a terrific add-on product called the "Video Toaster" that let Caleb edit home videos better than anybody else in the neighborhood. By 1992, when he'd finished his undergraduate work in design, he was looking forward to working with computers somewhere in the L.A. area.

Computer-related jobs in the L.A. area, even in those pre-Web days, were plentiful, and Caleb soon picked up work at a company that sold imaging workstations in Baldwin Park. The company was run by a couple of retired Navy guys who used their DOD contacts to sell souped-up PCs to other retired DOD guys running larger companies. They were sick of paying huge license fees to IBM for their minicomputers, which weren't much good in terms of "imaging" (scanning printed information into databases). Caleb liked the work but, within six months after being hired, the company fired everyone and went bankrupt. But Caleb didn't mope; he simply found another job, this one designing print ads for Red Dragon Systems, a Walnut Creek–based mail-order computer vendor that ran fancy ads in mail-order computer magazines such as *Computer Shopper*.

Red Dragon Systems, like so many fly-by-night computer companies based in L.A. in the early nineties, had a life span no longer than a California fruit fly, and it folded by the spring of 1993, leaving Caleb with a unique severance package: a flashy new 486 Graphics Workstation with graphics tablet, flatbed scanner, and a complete library of graphics software, including Photoshop, Quark, and Macromind Action, a predecessor of Director. To make ends meet, he got a steady, somewhat undemanding job at Fry's Electronics, but his real life was at night, squatting at his 486, creating spectacular images, animation sequences, and other graphics that, in time, would push the envelope of each software package he laid his hands on (often by "borrowing" a demo copy from the store, copying it, and returning it to the store shelves the next day).

Over the next five years, Caleb mastered just about every cyber-graphics tool, and when the Networked Economy dawned he felt ready to take his place in it. His job at Fry's hadn't been mentally rewarding; his salary was low and most of its customers were idiots. Nonetheless, the IQ of the grungy staff at Caleb's branch was stellar, and he spent many happy lunch hours picking the brains of the various gurus who ran the Hardware, Graphics, and Networking Departments of the store. At night he'd play with all the latest software tools: databases, spreadsheets, HTML tools, animation packages, and even the more obscure pearls of pleasure contained within alien operating systems, networking tools, and the like. Eventually, by process of osmosis, he had become what the New Economy needed—a Renaissance man—a human Swiss Army knife with a razor-sharp skill set.

The Dawn

In 1998, Web Fever hit L.A., giving the entertainment barons who controlled the metropolis an unwanted case of the jitters. Somehow (and nobody could figure how, because it had never warranted any coverage on *Entertainment Tonight*), the City of Angels had fallen far behind San Francisco, New York, and even such crude outposts of civilization as Seattle and Austin in the Great Web Revolution. To respond to this Godzilla-like threat, city councils were mobilized, New Media organizations hastily formed, and impressive budgets sketched on oak-paneled whiteboards. "We're not going to play second fiddle to anybody," said the Mayor. "We're ready to rock," said the newly appointed Interactive Director of a big L.A.-based record label. "Surf's Up" became the new battle cry in L.A., which quickly rebranded itself "The Digital Coast."

Soon, agents of Disney, DreamWorks, Paramount, Sony, Twentieth Century Fox, Universal, and Warner Bros. were combing every last nook and cranny of the L.A. basin, seeking out anybody who knew HTML, Photoshop, or Flash, and "had a dream to bring about the Grand Convergence"—an event destined to fuse PCs, video games, movies, and cable TV into an insanely profitable entertainment machine. The quest for convergence was so intense that moguls were teaming up with geeks on a wide assortment of broadband content plays including DEN, Icebox, and Pop.com, whose backers included such Hollywood heavyweights as Ron Howard, Jeffrey Katzenberg, and Steven Spielberg.

Representatives of these wild-eyed electro-zealots set up shop at all the local universities, from Harvey Mudd College to

UCLA. At the same time, the heads of interactive agencies in New York and San Francisco, whose olfactory senses were as highly evolved as that of any offshore shark, immediately opened up branch offices in L.A. to get a piece of the action. Sunset Boulevard echoed with construction hammers as West Coast offices for Razorfish, Organic, Viant, iXL, and other oddly named consultancies popped up overnight. There were also numerous homegrown Net start-ups that had emerged quicker than bad B movies: eToys, Citysearch, CarsDirect.com, Goto.com, and NetZero were just a few. In a manner of months, L.A.'s New Media industry, once a forlorn-looking go-cart, had become a fabulous muscle car ready to race down the nearest beach at two hundred miles per hour.

Caleb, still working at Fry's, was scooped up one day by a young, tanned, Ray-Ban–wearing, Porsche-driving recruiter who had gone into the store to buy a new Powerbook (his old one had spontaneously incinerated itself on the red-eye, which nearly brought the plane to an emergency landing over Kansas). He grabbed Caleb by the lapel, called him "Baby," and wouldn't leave until Caleb had agreed to meet with him—that week—in a plush office overlooking the La Brea Tar Pits. Leaving behind the routers and SCSI cables, Caleb quickly accepted a full-time position at LA Web, a design shop the recruiter had told him was "the hottest agency on the Web."

Go, Baby!

When Caleb was hired at LA Web, the company had just scored its first entertainment portal deal—a $300,000 job to build a network of sites for a major L.A. label specializing in punk, metal, and soft-core R&B. To the label, spending $300,000 wasn't a lot of money—not much more than shooting a music video for one of their acts. To Caleb and the other designers, $300,000 was a fortune, and they felt duty-bound to perform a first-class job. They did, working long, hard, and fast to build a compelling site replete with animations, sound loops, chat areas, CGI, and other "sexy" features.

This initial success (which subsequently won a Webby Award) led to a series of equally profitable assignments for LA Web: a half million dollars from Sony; $400,000 from Paramount; $600,000 from Warner Bros. Boom, boom, boom. In the front office the cash registers were ringing; in the back office Caleb supervised production like a Grand Master, spinning these multiple projects as adeptly as a pair of synchronized platters. If somebody had a problem or an issue, or if something on the site wasn't working, Caleb wouldn't just scold them and tell them to fix it; he'd work with them directly, rolling up his sleeves and playing with the code until it worked. For these efforts, Caleb was soon promoted from designer to "divinely lead designer and tech exponent"—a title whose humorous flourishes indicated that he was really the only person on staff who knew how to glue all the various pieces together and get the job done.

Hours were long, but to Caleb the work was fun—at least at the beginning—when deadlines were tight, and the only thing that seemed to matter was pleasing the client with a first-class job. But soon, as LA Web became more profitable, its Senior Managers concluded that they needed to "professionalize" its operation by hiring Claude, a former Art Director at a San Francisco magazine with a reputation for efficiency. Formerly trained in Print Production and fitfully unequipped with social skills, Claude would bark orders and get all hot and bothered if the product spec was messed with (even if the Web site that resulted was better as built than as specified). And, even when things didn't blow up, he'd gleefully relish any opportunity he had to bash Caleb's improvisational style with a lecture on "going through channels."

To Claude, running a tight ship was what his masters had hired him to do. To Caleb and everyone who worked under him, seeking approval from Claude for every microscopic project detail wasn't just insane, it was an affront to their artistic sensibilities. Just about everybody realized that if they were to obey him blindly, all ten of LA Web's carefully balanced projects—like pie plates spinning in the air—would soon come crashing down.

Caleb wasn't very good at outwitting a supervising fanatic, and he hated being forced to endure Claude's picky critiques of his work, especially when they were accompanied by screaming, which Claude seemed to resort to when he realized that his subordinates knew much more about the way Web sites actually worked than he did. One evening, after being called on the carpet over yet another insignificant design issue, Caleb had had enough, and became ready to think of moving on.

It wasn't just that Caleb believed that he could do better than the $80,000 that LA Web was paying him, or the fact that Claude was such a sadistic dolt. After working at LA Web for a year, and getting to know the company's clients, Caleb felt a little guilty about the whole "scam" the company was based on. They charged these people hundreds of thousands of dollars for Web sites that rarely cost LA Web more than ten or twenty thousand to build. This form of "Information Superhighway Robbery" was, of course, a well-known aspect of the Web consultancy biz. After all, if the client is a rich idiot, isn't one duty-bound to free him of his extra cash? And what could be more just than gouging an entertainment conglomerate whose main job was to gouge consumers with $20 CDs and $9.50 movie tickets?

Caleb was no moralist, but LA Web's brand of client-gouging was enough to make Hannibal Lecter lose his appetite. When Claude and his army of chirpy young female account execs were trying to land a client, it was all sweetness and light: dinners at expensive restaurants, energetic brainstorming sessions, PowerPoint presentations, tickets to sporting events—the works. Before the ink had dried on the contract, however, the attitude quickly changed to: "Fuck the client. They're clueless, anyway. They should be thankful we agreed to build their miserable site. Any problems, let the dogs in Production handle it."

While these shenanigans sickened Caleb, it was Claude's micromanagement that was driving him nuts. He began to dread coming into the office, and his innumerable daytime tasks were occasionally interrupted by bloody fantasies: Claude's head on a stake, Claude crushed under a server rack, Claude's mangled body being pulled from the wreckage of his brand-new Porsche Boxster.

During more rational moments, Caleb wondered whether there wasn't an alternative to such madness. Might he not be better off at another start-up? But after scanning several local job boards, and talking to other designers, he kept hearing the same old story: "They're all a bunch a bunch of lunatics. You think Razorfish is bad? Our Creative Director has personally fucked every designer in the office, male or female." Other designers reported equally horrific stories, such as the Web consultancy whose most important employee was its coke dealer, who kept the staff wired around the clock.

To Caleb, then, the question instead became: Could he avoid these abuses and make enough money—say $80,000 a year—to support himself? Getting such a salary at a Fortune 500 company was out of the question; they rarely paid Web designers more than $60,000 a year. On the other hand, if he went out on his own, and specialized in building lower-cost Web sites for smaller clients, he might just be able to make ends meet. One morning, while looking in the mirror, he noticed enough new gray hairs dotting his temples to decide to get the hell out.

His exit strategy was simple: Approach the best people he'd been working with and get them some freelance money on the side. Getting anybody to work full-time was out of the question. Many of his coworkers, despite their frustrations, were attached to LA Web with "golden handcuffs." Their reasons for staying there ran the gamut from blind faith ("I'm waiting for my options to vest") to blind hope ("They're going to get rid of Claude someday") to simple laziness ("I'm really comfortable here") to complete delusion ("I feel loyal to this company"). Despite these reservations, Caleb was able to convince several of them to work on a "per-project" basis, on nights and weekends. These included a good database/CGI/PERL coder, a front-end graphics person, and an all-around production grunt. By September, Caleb had rented a modest office in West Hollywood and had landed his first project, an $18,000 face-lift for a local independent film distributor.

Indie

Going solo was wonderful and scary at the same time. He was living on his own terms, making good money, and, most importantly, he felt like a craftsman who shaped projects from beginning to end. His biggest challenge was proving that a one-man shop could do everything that a "full service e-solutions provider" could. Clients were so used to paying enormous sums of money to crooked outfits like LA Web that his prices seemed too good to be true. Still, word-of-mouth publicity from former clients worked in his favor, and soon he had a stable roster of clients who paid between $20,000 and $30,000 for full-blown, Flash-enabled, end-to-end Web sites to promote their forthcoming entertainment ventures.

By March of 2000, he was billing about $75,000 a month, and after paying his freelancers and other business costs, keeping about $20K—enough money for him to buy a new car—a silver Acura—and upgrade his own image. Soon, he had accumulated the same odd-looking assemblage of mobile gear worn by top-notch designers everywhere: cell phone, Palm Pilot, Blackberry, Mobile MP3 Player, and other assorted gizmos that showed technophobic clients that he was playing in the big leagues.

Keeping costs low meant that Caleb could laugh at the world, at VCs who came proffering money, or clueless agents of big-name media companies who wanted to buy his outfit for a million dollars and bring it in-house. He'd learned well from the big agencies how to move like a shark—always swimming with one's mouth open—but he was nimbler, more efficient, less greedy, and more inspired and responsive. And so he'd often sit back in his Spartan office and laugh. He had won what so many never have in life, a ship of their own and an ocean whose end if end there was, would deliver him to better land.

Survivor

Caleb's experience at LA Web had made him an early cynic about the promise of the New Economy. But when the Bubble burst in April of 2001, he was as surprised as the most wide-eyed booster. His initial reaction, forged by watching too much CNBC, was a fear of financial devastation: no more moon shot IPOs; no more unprofitable e-tailers trading at $300 a share; and, most threatening to his own bottom line, no more clients looking to outsource Web projects.

Although the epicenter of this disaster was located a few hundred miles north of L.A.—up in Palo Alto—aftershocks were immediately felt in the heart of L.A.'s Web scene. Among the companies feeling the shakes were Stamps.com, CarsDirect.com, CitySearch, and CDNow, all of which reacted by laying off hundreds of employees and abandoning their plans for world domination. Web consultancies like LA Web saw their client lists immediately cut in half, with those remaining clients too cash-strapped to pay their exorbitant rates. Many resorted to midnight mergers, last-minute restructurings, and reverse stock splits to stay alive.

Ironically, Caleb's fears of disaster proved unfounded. The shakeout crimped budgets across the board, making the concept of the $300,000 Web site seem as fashionable as the mullet. "The new, new thing" was "lean and mean," and you didn't get leaner and meaner than Caleb's one-person shop. Any lingering doubts about Caleb's ability to provide clients "more bang for the buck" went out the window. Soon, his phone was ringing off the hook, and he became so busy that he began drawing upon the talents of many of his former LA Web colleagues, most of whom were now unemployed and more than willing to freelance for him. During this frenetic period, Caleb was billing more than $200,000 a month and clearing $15,000 for himself. His workload was so intense that his office became his bedroom and his desk disappeared under a growing pile of invoices, pizza boxes, contracts, and miscellaneous receipts.

But like the New Economy, Caleb's run of good luck was not built to last. He went from ten clients in June to six clients in July. By the fall, he didn't know where his next dollar was coming from.

August to October

Caleb was scrambling for any possible project, no matter how small. Telephone book in hand, he cold-called local retail stores, small record labels, talent agencies—anyone who might be a potential client. The results were mostly dismal: People either cursed him out for being a telemarketer (which technically he was) or, even if they were interested in his services, they gave him the roughest time imaginable. These low-end clients wanted everything under the sun for less than a thousand dollars, and when he tried to explain the time and effort that went into building an entire e-commerce site, for instance, their attitude was, "I can throw a rock out my window and hit somebody who does this stuff. Take it or leave it."

Although he wanted to tell such people to go to hell, he did take it. He cranked out porn banners, a Web site for his dentist, more porn banners, a database for an off-shore electronic casino, even a home page for an old lady from the supermarket who wanted her grandchildren in North Carolina to see the latest pictures of her cat. As if the projects and the piss-poor rates he was charging weren't bad enough, he often had trouble getting the clients to pay. Since most of them were shady at best, and knew the amount they owed him wasn't worth a lawyer's time, they would laughingly say, "So sue us," or else make up bullshit excuses like, "Your work was substandard," then turn around and use it.

By October 2000, things got so desperate that he had to give up his studio in Manhattan Beach and move to a crappy apartment in Costa Mesa, just a few minutes from his parents' house. At night, he'd drive up and down Sunset Boulevard, speculating wildly about whether a given strip club had a full-featured, Flash-animated Web site that they might pay him more than $500 to build. During the day, he'd sit in his office, fielding phone calls from creditors, leasing agents, and telemarketers. Even though there was no new business, he tried to stay occupied, going through stacks of letters that he'd once been too busy to bother with.

His search yielded several small checks that had somehow fallen between the cracks. But it also produced an ominous official-looking letter from the IRS, dated several months earlier, that warned him that his quarterly self-employment taxes were already in excess of $20,000. Frantically grabbing the telephone book, his sweaty finger dialed up the first accountant he saw: Abe Abromowitz, of AAA Accountants.

"You've got to help me," Caleb cried into the phone. "They're dusting off the hot seat for me."

Epilogue

After meeting with the accountant, Caleb realized the unthinkable—that there was no way that he could remain an independent and pay his outstanding bill to the government. Using his calculator, the accountant quickly determined he'd have to build at least eighty-seven Front Page-grade sites at $500 a pop per month to make enough money to pay Uncle Sam. Doing so would also mean giving up such unnecessary expenses as food and shelter.

After several weeks of unsuccessfully looking high and low, his luck changed when a former freelancer of his tipped him off about a designer position in L.A.'s last remaining recession-proof industry: pornography.

Today, Caleb works in a diverse office in North Hollywood containing a state-of-the-art production Web facility that is staffed by several LA Web alums, along with a wide variety of silicone-enhanced models and the owner's pet cheetah. He makes less than $50,000 a year, but work is steady, the hours are predictable, and he's slowly paying off his tax bill.

— Vigilantes: Screaming for Justice —

Vigilantes: Who Are They?

When the New Economy tanked, everybody was caught in the shit storm. CEOs, Venture Capitalists, Wall Street Analysts, Bankers, and the Media were all accused of being complicit in the Ponzi scheme. The CEOs blamed the Venture Capitalists, the Venture Capitalists blamed the Wall Street Analysts, the Analysts blamed the Bankers, and the Bankers blamed the Media. This never-ending circle jerk of blame became a popular parlor game for everyone who hated the New Economy in the first place.

But for people who had been truly screwed, the blame game, the buck-passing, the circle jerk, was not enough. These frustrated souls, seeking Justice at any price, became Vigilantes who fought back using any means at their disposal.

Some became Vigilantes overnight, when they were fired without cause, notice, or severance. In Seattle, at the office of Web Design company MarchFIRST, employees reporting for work found themselves locked out and responded by breaking windows, vandalizing property, and stealing whatever they could carry from the building.

Others, still employed but disgusted, became Vigilantes over a matter of months. On the outside they were still the optimistic, energetic dot.commers working sixty-hour weeks with nary a complaint. But when the moon was full and the bile was rising in their gullets, they fought back by leaking rumors, embarrassing memos, and other sensitive data to the media, especially to Web sites such as Dotcomscoop, FuckedCompany, and other public bulletin boards. These sites encouraged the digitally disgruntled from across the globe to add their dirty little secrets to the stinking pile of evidence.

Beyond violence and whistle-blowing, lawsuits became a favorite method of attack. There were wrongful termination suits, sexual harassment suits, and breach of contract suits. But working within the system had its limits. Vigilantes had no use for unions, which had already been crushed at Amazon and thwarted at e-town by early 2001. The libertarian mindset of Vigilantes has no room for collective action. They confronted the New Economy's ills by themselves, like gunslingers at high noon. Eschewing a pair of Colt 45s, these angry men tried as best they could to defeat the forces of New Economy Evils with a mouse, a keyboard, and a monitor.

What gives a Vigilante satisfaction? To see every CEO whose face appeared on the cover of *The Industry Standard* strung up by his thumbs? To see Mary Meeker, Henry Blodgett, and every bullish analyst tarred and feathered with their own ridiculous "buy recommendations?" Or simply to see what's already happened: The New Economy turned into a twenty-first-century Boot Hill?

If our experience with Vigilantes is any guide, we suspect that nothing short of Divine Retribution would suffice.

Are You a Vigilante?

You might be a Vigilante if...

- You would like nothing better than to run into Stephan Paternot in a dark alley. (Or if not theglobe.com's egotistical cofounder, you'd be willing to substitute Henry Blodgett or Mary Meeker.)

- Your version of a severance package from your dot.com job consisted of all the brand-new Sony VAIOs, flat-screen monitors, and assorted hardware you could sneak out the back door.

- You've customized "Half Life" and "Counter-Strike," using a floor plan of your last company's offices.

- The same backpack that used to be filled with *The Industry Standard* and O'Reilly books is now weighted down with dossiers and "evidentiary proof" you fished out of the executive washroom.

- Your friends and family have learned not to ask you "how are things going?" for fear of having to endure a thirty-minute non-stop tirade.

- You just posted your ten thousandth message on FuckedCompany.com. (Pud, needless to say, is your hero.)

- The amount of money you lost investing in a dot.com is less than what you currently owe your attorney.

- You refuse to admit that you ever believed that stock options were your ticket to early retirement. After all, extreme negativity is merely an overcompensation for blind positivity.

Fun Facts about Vigilantes

How They View Themselves: Charles Bronson, Steven Seagal, John Wayne, Clint Eastwood, Patrick Swayze (just kidding—we wanted to make sure you were paying attention), Bruce Willis.

Favorite Pastimes (During the Boom): Mountain biking, snowboarding, and aggressive foosball.

Favorite Pastimes (After the Bust): Soul-searching, dumpster-diving, and rumormongering.

How Other People View Them: Long-winded, whiny pains in the ass or scarily angry and obsessed maniacs.

Psychological Profile: 95 percent rage, 5 percent self-hatred.

What They Did during the Web Boom: There was no specific or typical job for Vigilantes. They could have been anyone. However, it does seem like a rule of thumb that Vigilantes were located either way down on the totem pole (people who didn't receive severance) or in its uppermost reaches (which is to say that you're owed millions from the stock options you were promised, har, har).

What They Did After the Boom: Bitched, moaned, broke things.

Current Employment Situation: Inconclusive. Since Vigilantes came from every level of the tech biz, some are probably working and some aren't. But either way, they still haven't given up their revenge fantasies. (Like us, for instance, writing this frickin' book.)

Vigilantes: The Story of Vincent

None of the staff at CRT Wireless was prepared for what they now describe as the "single most humiliating experience" of their working lives. Around 2:30 on a Wednesday afternoon in January 2000, when management was confident that the start-up's ninety various salespeople, developers, and administrative personnel had all returned from lunch, a company-wide e-mail was sent around the office directing each recipient to report to the conference room at a specific time. The tone of the missive was ominous, yet things had been so good at CRT that they were totally blindsided by the events that were about to unfold.

At one end of a long conference room table sat a mini-delegation consisting of the head of HR, the CFO, and two nameless, gray-suited men no one had ever seen before. Thinking that the meeting probably had something to do with stock options, many of the employees strolled into the room with big smiles on their faces, expecting good news about how rich they soon were going to become.

What followed was nothing less than what old police shows used to call "The Third Degree." While it wasn't conducted with electrodes, truncheons, or cigarette burns, there were enough veiled threats involved to make its effects equally devastating to everyone who went through it.

"What do you know?" asked one of the gray-suited men, who turned out to be security consultants, "and when did you know it?"

"Know about what?" asked the hapless, helpless employee, nearly blinded by a Tensor lamp pointed at him from a nearby credenza.

"Aha," said the other consultant. "You DO know something, don't you?"

"Let's not play games," said the HR Director, clearly weary of this sort of pointless back-and-forth. "Look," she'd ask the candidate. "We're not after you. In fact, you can use this opportunity to make it easier for yourself. Just tell us this: What do you know about the parser?"

"'Parser?' . . . What's that?"

"We'll ask the questions," snapped the first consultant. "Now, what do you know about the parser?"

"You can save yourself a lot of trouble," said the HR Director, "if you stop playing dumb and tell us what you know."

Few of the ninety people who survived the surreal grilling had much of a clue about what its ultimate objective was, but by the time their fifteen minutes in the hot seat were up, they were allowed to leave, seemingly unscathed. While most felt lucky that they hadn't been immediately transferred to a prison cell replete with chains, snakes, spikes, spiders, and rats, for the remainder, their emergence from the windowless room did involve one further humiliation.

Returning to their cubes, they were met by a uniformed security guard who told them to pack up their personal effects immediately. Everything else, classified as "company property," had already been removed. This included computers, monitors, laptops, Palm Pilots—any technological device in which a "parser" could possibly be concealed. It was a scene repeated over and over that afternoon. The relief of having survived the interrogation turned to shock at seeing the security guards and, finally, anger when the spectacle of the stripped-down cubes meant that nothing they could've said in that room could've saved them. They were screwed even before going in.

By 4:30, approximately thirty employees had been put on "indefinite administrative leave," pending further investigation. To those affected, it seemed like a tidy way of getting rid of them. At least that's the way it felt, after they had been escorted through the glass doors, down the staircase, and into the parking lot, all under the watchful eye of the guards.

In any case, it was an odd and unexpected turn of events, considering that only a few months earlier CRT Wireless had filed its S-1 in eager anticipation that their suite of wireless Internet applications would make for an incredibly successful IPO.

Confused

That night, at a T.G.I. Friday's across town, an angry mob of former and shaken employees gathered to toss back a few and hopefully figure out what the hell was going on. After five pitchers of beer had been emptied in quick succession, the truth, mixed with wild speculations, began to emerge from around the table, upon which a single candle flickered.

"It's got to be Vincent's fault," said a beefy developer named Hank. "Everything points in his direction. The business about the parser; the head Nazi kept going back to that."

"What the hell *is* a parser?" asked Sheena, an administrative assistant wondering about how she was now going to feed her two kids.

"I don't know what they hell they meant," said Hank. "But 'parser' is just a generic term. It's just a part of a program that takes input, checks it, and spits something else out. Like in a Web browser. You do know what a Web browser is?"

"I use AOL," said the administrative assistant.

"Well," said Hank, pouring another beer, "I should have expected that."

"Why would Vincent be working on a parser?" interrupted Vladimir, a junior developer. "He just makes those stupid demos for the sales people."

"I don't think those fucks had a clue," said Hank. "They were on a fishing expedition. But we're probably not ever going to know about why we lost our jobs today until Vincent shows up."

"He hasn't been around for a week," said Sheena. "Where could he be?"

"Probably Mexico," said Hank. "Or Canada. Or maybe in hiding."

"Unlikely," said a voice behind the table. It emanated from a large-framed man who had suddenly loomed from behind the bar. In the dim light he looked a bit like a young Orson Welles.

"Vincent!" said Sheena. "Won't you sit down?"

"Don't mind if I do," he said, as chairs squeaked across the floor to accommodate his wide frame.

Red-eyed, drained of blood, all faces turned to the man who'd just sat down, asking the same unspoken question: "What the Fuck Happened?" Vincent just sat there, with a faint smile on his face. In the near-silence, marred only by the faint, echoing dance beat of the Bee Gees, he milked the dramatic moment for all it was worth, his eyes passing around the table, until he was sure each face was at rapt attention. Then he began to speak, in a low voice.

"What I am about to tell you can never pass beyond this table. Do I have your solemn assurance?" Each face nodded. "All right, then. Prepare yourselves for the damnedest story you've ever heard."

The Wonderful World of Wireless

Vincent was in his mid-twenties when he was hired by CRT Wireless in 1998, succumbing to the siren call of "a dynamic work environment," "a liberal compensation package," and naturally "stock options," which would come his way when the company did its public offering. Although he was a skilled developer, he was hired for one purpose only—to put together "vaporware demonstrations" for CRT's sales force. The demos were custom-tailored to sell its clients on the glorious potential of "pervasive wireless computing."

Many programmers might have balked at the prospect of devoting their lives to crafting such elaborate fakery, even if it meant $85,000 a year and a shot at "the big-time money." To Vincent, however, "vaporware" was an essential part of the business. It was something everybody did, from Microsoft to Oracle on down. It wasn't evil or deceptive; it was just "concept marketing" and, in the New Economy, concepts often mattered far more than nuts-and-bolts code, whether it was the myth of e-commerce, the "killer app," or, in the case of CRT, the fabulous possibilities of the Wireless Internet.

Like many other "great ideas" that were born during the dot.com bubble, CRT's Wonderful World of Wireless was based on the reasonable notion that people—wherever they were—would crave remote access to all of the bits and pieces of the "information space" necessary to orient them on the planet. In this Brave New World, FedEx trucks cruised by "Smart Drops" that instantly told them how many packages were in the drop, their weights, even their destinations. Wirelessly enabled business people, before boarding their planes, could send information ahead to their hotels, preordering their room service dinners, preprogramming their cable channels, and preordering every possible form of after-hours entertainment, from after-dinner mints to Altoid-sucking hookers. The possibilities for the wireless world were so endlessly personalizable—and presumably profitable—as to defy imagination.

The problem, of course, was that none of these futuristic schemes actually worked very well in practice. Although subscribers in Japan and Sweden were often able to use their cell phones and other mobile hardware for tasks more sophisticated than simple gabbing, U.S. consumers were not so lucky. After laying down $400 or $500 for even the most cutting-edge, "wirelessly enabled" PDAs and cell phones, they often found themselves disappointed to learn that their magical devices rarely functioned as advertised. Unless one loaded oneself down with $6,000 of gear from the likes of Palm, Visor, Compaq, Blackberry, Nokia, and others, getting access to more than a few blessed fragments of the glorious "Information Space" was impossible. Often, you could barely get a signal even if you happened to live within 500 meters of a cell transceiver or a low-orbiting satellite. Whether the blame for this sorry state of affairs was to be laid at the door of the FCC, the Baby Bells, the ITU, or some other bureaucracy kept legions of beard-stroking Info-Pundits amused, but not consumers, who, after experimenting with these fancy devices for a few weeks, often handed them down to their kids, who fooled around with them for a few days before retiring them to their toy chests.

Fortunately for CRT and other wireless companies, this mass realization of how truly primitive most devices were would be slow in coming and, in the meantime, wireless was hailed as the "new, new thing" that was going to save the technology biz. As interest in e-commerce was waning, the stock of wireless hardware and software makers was soaring. Motorola, Ericsson, Nokia, Palm, and Handspring—there were just a few of the outfits riding high on the astonishing growth rates, which reminded many of the PC Revolution of the early nineties.

To be fair, it wasn't just investors and consumers who were hoodwinked by such science fiction. Many telcos (telephone companies) eager to keep pace with the competition, spent billions in 1999 on 3G spectrum licenses only to realize six months later that they had bet the farm on a technology that was years away from reaching the market.

The Emperor Has No Connectivity

Hardened cyber-cynics might have concluded that only an industry-wide conspiracy to foist flaky technology on the public could account for this ridiculous state of affairs. After all, wasn't the New Economy nothing more than one big Ponzi scheme? Vincent, watching these grand maneuvers from his tiny, back-office perch, could only note that in CRT's case, the primary goal was to present a fabulous illusion that obscured the fact that its own technology was worthless.

Seated in the midst of the real developers, his inbox overflowed every day with fevered requests from sweaty salespeople chasing down Fortune 50 clients in elevators, trade shows, cocktail lounges, and strip joints all across America.

For phone companies, he would mock up elaborate voice-to-text messaging products; for pharmaceutical companies, he would conjure up visions showing how reps in the field could wirelessly transfer leads and other sensitive data; for ISPs, he would present elegant wireless Instant Messaging solutions or friction-free ECRM utopias for struggling retailers. There was no end to these Tomorrowlands, these hands-free nirvanas delivering wireless ecstasy on demand.

He used Javascript, XML, and Flash to knock the demos out as quickly as possible. His only distraction came from the screams and curses that often came from his cube mates, the developers, struggling to get the underlying architecture to perform basic tasks.

"Mother fucking shit," one developer screamed in a heavy Russian accent. "Who wrote this? The technology must be six years old!"

"This Trunk Concentrator is strung together with chicken wire," groused another. Such complaints weren't just the bitchings of egotistical programmers. The product really was a dog. At bottom, CRT's vaunted Enterprise Gateway

Technology was an antiquated messaging suite written in DOS that had simply been overlaid with an NT shell. No one in his right mind would ever believe that it could work. But then, most clients weren't in their right minds. Greed, wishful thinking, and ridiculous industry projections, along with Vincent's elaborate and convincing demos, had blinded their judgment.

Among those who had been taken in was none other than AOL, a company that prided itself on avoiding dot.com insanity, wired or unwired. The Virginia-based giant, seeking to brand its evil pyramid icon on every mobile device in the world, signed a letter of intent for CRT to develop a new Instant Messaging product for its 30 million subscribers, and went so far as to invest $10 million in the company.

The mood inside CRT's drab suburban offices turned hysterically triumphant. It was as if everyone had won the lottery. Imagining themselves but a few steps away from the pot of gold at the end of the Internet Rainbow, champagne corks popped, high-fives slapped through the hallways, and the parking lot—within less than a week—was populated with BMW Z3 roadsters, Audi TT Fastbacks, and SUVs so large that they looked like houses with wheels.

Vincent half-heartedly took part in the riotous, weeklong revelry. But, even before the last bottles of champagne were emptied, he began to feel a creeping anxiety that wouldn't go away.

"What are we going to do when they figure out that we've got nothing?" Vincent asked one of the sales guys.

"Don't worry. We can build this product fast. You'll see. We'll have it launched before anyone's the wiser."

This response, while meant to assure, disturbed Vincent even more. He knew that unless God Himself was hired to perform a coding miracle, there was no way that the product would ever do a tenth of what the salesmen claimed it was already doing. Equally disconcerting to Vincent was the sight of limousines in the parking lot that had ferried investment bankers out to the wilds of Long Island. Bear Sterns, CFSB, J.P. Morgan—firms that had not previously returned CRT's calls—were now sending their best and brightest representatives, all clutching the same *Wall Street Journal* article announcing AOL's investment.

To Wall Street, CRT had become an overnight sensation. Not a day went by that the main conference room wasn't overflowing with immaculately dressed Wall Streeters, talking up the "global execution capabilities" of their team, the "commitment of analysts," and their firms' track record on technology issues.

"Jesus Christ," Vincent said to himself, "What have I done?"

That Sunday, as usual, he visited his parents for dinner, and was still so upset at what was happening at work that he tried discussing it with them. Aside from his cousin, who did tech support for Computer Associates in Islandia, New York, his family had no idea what he was talking about. All they knew was that he was a good kid who worked on computers and wasn't a "mamaluke" like his brother Joseph. His father, who owned his own construction business, and bore a striking resemblance to Tony Soprano, was especially clueless about computers, and was unwilling to take Vincent's problems seriously.

"You make a lot of money with that computer crap. Who cares if it's real or not?"

"But, pop, what if Mr. Cipola asked you to build a giant fountain in his backyard and you knew there was no way in hell you could do it?"

"Mr. Cipola?! That cheap bastard would never hire anyone to do nothing. They're gonna bury that fuck with his money."

"But what if someone else, someone who isn't cheap, wanted you to build something you couldn't build? What would you do then?"

"I'd get some Mexicans to do it. Those guys work like hell and you don't have to pay them too much."

"Pop, it's not that simple . . . "

"Oh, how should I know?! You're the computer genius. Isn't there some Poindexter Guy someplace who could help you?"

Enter Sandman

In truth, no one could help CRT Wireless. And the louder the developers complained, the more they were ignored. The company seemed gripped by a case of Mass Denial, manifested by a complete unwillingness to face the facts. This behavior included everyone from the lowliest admin to the top executives. Chief among them, in every respect, was the president of the company, Al. Al was an affable guy in his sixties whose prior claim to fame had been selling network in the early seventies heyday of mainframe computing. His idea of management was to parade through the hallways, shaking everyone's hand and slapping them on the back.

"Nice to meet you, Gregory. Great job you're doing!"

"Steven, the new Web site looks great!"

No matter how much he dealt with Al, Vincent couldn't seem to get used to him because Al could never remember anybody's name. "I've met him three times before," complained Vincent to his boss Victor, "and he's mixing me up with someone named Gregory. I don't think that there's a single Gregory working here."

"He's a kook," said Victor. "That's the only way I can explain it. I've been here three years and I think he's always mixing me up with some guy he used to know in Korea. When he was interviewing me, we went out to dinner and he started talking about how his father used to beat him in their Lower East Side tenement, and then he started crying."

"You're kidding me."

"I wish I were. Once, during a sales meeting, he had us up doing yoga and told us to hold all our tension in our crotches."

"I sometimes wonder about where this whole thing is leading," said Vincent. "Don't be deceived. Al's crazy, but he's crazy like a fox."

It was only after the Bubble burst in April of 2000 that Vincent began to appreciate just how wily Al really was. The market slowdown eliminated the possibility of Initial Public Offerings for many companies; the result was a wave of mergers and hostile takeovers among formerly fierce competitors. Palm's successful IPO was a thing of the past, and its stock had since cratered. Even AT&T's offering proved lackluster, leaving everyone—from giant Vodafone to the smallest Baby Bells—seeking protection in the arms of corporate sugar daddies. Now the phrase that paid wasn't "IPO," but "M&A." Merger-mania also hit the wireless applications provider space. Phone.com snapped up rival Software.com, and rumor had it that Aether systems—whose stock was still hovering at a high $70 price per share—was looking for acquisition targets.

With only $5 million of venture capital left in its coffers, CRT wireless went all out to make an impressive showing at Internet World in Chicago. They bought the largest booth space available, strategically placing themselves next to Aether's booth, hoping to strike up a conversation, perhaps more. Amidst the din of booth babes, mimes, skateboard competitions, and chattering wireless evangelists, Al and the team descended on Aether's booth like a squad of paunchy commandos. Some were instructed to gather up all the sales material they could grab, others were encouraged to ask technically daunting questions that would impress Aether's tech reps with CRT's expertise. Vincent, hundreds of miles away, was told in no uncertain terms not to leave his workstation for the duration of the show, for fear that something might go wrong with the demos.

In the meantime, Al put his glad-handing skills to work. Knowing that Aether's CEO was a helicopter enthusiast, Al engaged him in a lengthy conversation about the Sikorskys he used to fly in Korea. As strange as the whole pitch seemed, it worked—by the end of the show, Aether was eating out of CRT's hand, and a deal seemed imminent.

The hook that Al played up to the max was how Aether was strong on applications, but weak on hardware. "Look, you've got the fuselage and we've got the rotor. Together we make for a great combat chopper."

Al was able to close the Aether deal very quickly. The announced purchase price was $120 million, which broke down to $20 million in cash and $100 million in Aether stock.

Perhaps more notable than the deal itself was Al's severance package. Instead of loading up on Aether stock (which was in the seventies at the time), he asked for a cool million in cash and promised to go away—presumably, to the nearest golf course. Everyone, including his executives, thought he was nuts (okay, completely nuts) for selling himself short, but Al was from the Old School, where a pile of money in the hand was worth more than a zillion options in the bush.

Al's eschewing of virtual wealth didn't stop him, however, from approaching the wives of several employees at the Christmas party a few weeks later, saying, "How does it feel to be married to a multimillionaire?"

On the Side

Vincent, because of his low-end position, wasn't about to become a millionaire. While his 1,000 options in the combined company would net him at least $70,000 (as long as Aether's stock didn't totally tank), he wanted more from life than virtual wealth from a virtual product. While other people might have rested on their laurels, thinking that they had hit the jackpot, Vincent began working busily on his own secret project. His motivation was part boredom, part guilt, and part ambition; in short, he wanted to create something that he would be proud to put his name on.

For months, he labored, night after night, hunched over his home computer until the wee hours. Teaching himself how the dazzling cornucopia of wireless middleware worked (or didn't work), he started playing around, using the Java programming tools that he knew, some C++ and XML, and a little PERL. His coding was messy at first, but every line he wrote was inspired. He knew that he was reaching, but he knew what he was reaching for—something that might take all the wireless chaos and tame it in a way that could make the end user's life a little easier. One night, his fumblings finally came together, congealing like a tasty omelette to the point that order was imposed on chaos. After a few weeks of bug testing and informal USENET peer review, he brought it into the office to show to Victor, his boss. Victor was impressed enough to arrange an audience with John, the CTO.

"Gee, Vincent," John said, after spending about an hour fooling around with Vincent's handiwork, "We can really use this. I'm not sure what we can use it for, but it might just be a suitable platform for other applications. What do you call it again?"

"The Universal Parser," said Vincent.

"If I just ZIP your file and send it to engineering, will they be able to plug it into whatever their hot little hands are working on?"

"I'm afraid not," said Vincent. "It's compiled, so they're not really going to be able to plug it in. But I've got the source code at home. I can bring it in tomorrow, if you think it's worthy of their attention."

"Good enough, man," said the CTO. "I can't tell you how hot I think this is."

"Really?"

"Really, really. You know, we might even be able to build it into one of our proprietary protocols and call it 'The Smart Protocol,' or something sexy like that."

"That's great."

The next day, Vincent woke up late, after spending half the night coding, and forgot to bring the source code. But ultimately it didn't matter because John was nowhere to be seen. The office was in a complete uproar that morning because, due to some critical communication error, the demo for the wireless Net-enabled refrigerator got mixed up with the wirelessly enabled soda machine, causing one of their top sales guys to look like an asshole in front of the client.

Vincent laughed off the situation, fully assured that his ascent in the organization would only have to wait a day or so until everything settled down. In retrospect, he couldn't have been more mistaken.

Bucket of Bolts

By early January, a team of Aether people had flown up from Maryland and had installed themselves at CRT's Plainview headquarters. Although they claimed to be there "to learn how CRT did business," it soon became obvious that their true mission was to determine how many people to lay off.

Being adept at denial, none of CRT's staff picked up on this fact, or at least allowed themselves to be consciously aware of it. Instead, they took Aether's statements at face value and eagerly showed them whatever they were doing. For the first time, Aether's top engineers were looking beyond the bold promises, and began scrutinizing what CRT had under the hood.

These inspections prompted many closed-door meetings, and conference calls that devolved into shouting matches. The climax of this conflict came in late January, when all of senior management hunkered down for a twelve-hour session designed to resolve any "outstanding issues" related to the viability of the deal going forward.

By 6:00 P.M., the conference room table was a mass of crossed-out technical specifications, coffee-ringed product documents, and half-eaten sandwiches discarded on tiny Styrofoam plates. The once-pristine whiteboard was now a mass of dotted lines, and circles and arrows leading to certain oblivion.

Conversations that had begun in hushed and civil tones had degenerated into hoarse, angry epithets. Only a handful of people knew what was really going on, and they made plans to leave work early that evening to avoid any direct repercussions.

Vincent, hunched in his cube, suspected the worst, and worked as slowly as possible, in hope of witnessing the outcome. At 7:45, the door opened, and the two senior Aether executives made a beeline for the water cooler, which was only six feet away from Vincent's cube.

"God, I'm thirsty. Screaming dries me out something awful," said one Aether exec to the other, filling his Dixie cup from the plastic jug.

"This company is a fucking joke," the other said. "In my fifteen years in this industry, I've never seen such crummy software."

"Look, I know it's a bucket of bolts, but that Universal Parsar thing, we could make a billion dollars with that thing. A whole fucking new product line."

"Well, let's take a look at the source code before we cream our jeans."

Vincent sank down low in his chair, so his head wouldn't be seen. The men soon returned to the conference room, and when Vincent was sure he was safe, he crept out into the night.

That night, Vincent dreamt that his humble little widget had become the Microsoft Office of Wireless Middleware. He saw himself in Comdex, in front of a thousand cheering wireless zealots, as he gracefully accepted an award from Craig McCaw. The booth babes he had once leered at from across the trade show floor were now falling over one another to hand him their hotel keys. Descending the stage, he found himself mobbed by hundreds of reporters clutching notebooks, eager to record his every utterance. It was as close to a wet dream as he had had in quite a long time.

The Big Squeeze

At 9:30 the next morning, after a peaceful night that had left him feeling bright-eyed and rested, Vincent was happily reading through his e-mail when John the CTO peaked over his shoulder.

"Hey, Vincent, could we have a word with you for a minute?"

"Sure," he said. Within moments, with a spring in his step, he made his way to that same conference room that, only a few hours before, had been the scene of such heavy discussions.

Seated around the conference table were John, Victor, and Aether's main integration engineer. Also there were Aether's corporate counsel and HR Director—whom Vincent had never seen before.

"How's everybody today," Vincent beamed, offering his hand to the assembled crowd, which quickly sat down. The integration guy opened up a manila folder and pulled out what looked like a twenty-page memo.

"So we understand that you've been working on something quite exciting. You call it a parser?"

"My parser," Vincent said, "it manages to speak all of the various languages and protocols very well. In fact, you might say that its robust core . . . "

"We know what it does, Vincent," said the CTO. "The reason that we're here is that we'd like to clarify your position in relation to this project."

"Okay," said Vincent.

With that, the integration guy slid the document across the table. Vincent opened up the first page and squinted at the mass of legalese, printed in tiny, 6-point type. He couldn't make out what the document actually said.

"What's this?" he asked.

"It's a simple agreement calling for you to tender some rights and ownership of your application to the combined organization."

Vincent nodded his head, before nervously asking, "What do I get out of it?"

"For one thing," said the HR woman, "you get the right to keep your job."

"I don't understand," said Vincent. "I thought you guys liked my product."

"Our product," said the integrator guy.

"We're prepared," said another Aether executive, "to offer you a very, very sweet deal. What would you say to a $2,000 salary increase, in addition to 500 more stock options?"

Vincent, remembering the conversation he had overheard the night before, did his best to make up an excuse.

"I'd really like to show this to an attorney before signing anything."

"You know," said the corporate counsel, "that you developed this on company time, and according to your employment agreement, it really belongs to us anyway. We're really being unusually beneficent here."

"Don't get me wrong," Vincent said, now realizing what he was up against. "I really appreciate your generous offer. Would it be okay if I waited until the end of the day to turn in this executed document and deliver the source code?"

"By all means," said the corporate counsel. "Just as long as it's by the end of the day."

Flight and Fight

Vincent picked up the document and walked slowly out of the conference room. But, as soon as he was sure that no one was looking, he made a mad dash for the parking lot. "They're trying to fuck me," he said to himself as he jumped in his car and screeched out of the parking lot in a cloud of blue smoke.

He arrived home in record time, despite the usual congestion of midday Long Island traffic. Without even stopping to shut the door behind him, he raced up to his computer room to see if his application was still there. Furtively typing his login password, his worst suspicions were soon confirmed. At 9:47 that morning—just about the same time that the integration guy had slid the paper across the table—somebody had breached the firewall on his home network and had copied his entire hard drive.

"Those fucking bastards," said Vincent, but instead of feeling raped or otherwise violated, he felt triumphant, remembering that the only place that the parser source code actually existed was on a floppy disk that had been stuffed into his shirt pocket throughout the whole meeting. He tapped the disk to make sure it was still there (it was), and reflected that if the integration guy had simply reached over the table, grabbed him, and shaken him upside down, he'd have been spared the trouble of having someone hack into his holiest of holies.

The next day, Vincent called in sick—something he'd never done before—drove over to a lawyer's office, and told him the whole story. After finding out that most case law supported his position—which was that unless an employee specifically signs an employment agreement, he owns all "inventions" that he develops on his own time, Vincent put the lawyer on retainer, and told him to begin work on a lawsuit alleging trespass, attempted theft of property, and intentional infliction of emotional distress.

On the advice of the lawyer, he also filed a complaint with the FBI's Cyber-Surveillance Office. Before retiring that night, he changed all his locks and installed a motion detector alarm he'd bought from RadioShack on the way back from the lawyer.

No Justice, No Peace

Vincent finished the story, and Hank refilled his glass. It was getting late and the group was in a funk. They didn't know whether to admire Vincent for outwitting CRT's thuggish managers, or string him up on the nearest pole for costing all of them their jobs.

"Why didn't you just turn that shit over to them?" asked Hank. "They probably would have paid you anything you wanted. You had them over a barrel, man. Now we're all fucking unemployed—the company is as good as dead—you're probably even afraid to go home. I know I'd be."

"I'm sorry," said Vincent. "Maybe I can't explain it. But when you work hard at something, and it's yours, and they try to steal it from you—deny it's yours—breach your firewall, and fuck with your hard drive, well, it just pushed the last button. The one that says, in big bright lights, 'I don't care anymore.' They pushed it—and you can't reset that button—it just stays pushed in."

"What are you going to do?" asked Sheena.

"Disappear for awhile. I can't really leave the country—not as long as this lawsuit's on. Maybe I can do some coding work. I'm sure there's something that I can crank out that somebody will pay for."

"What happened to the parser?" asked Hank.

"You mean this?" answered Vincent, reaching into his shirt pocket and producing a 3½-inch floppy disk.

"Yeah," said Hank, looking at the disk as if it were the Hope Diamond.

Vincent cracked a half-smile and slowly, ever so slowly, moved the disk above the flickering candle. The heat began to distort the shell, scorching the label, melting the disk inside.

"My God," said Sheena. "What are you doing?" The disk, slightly flaming, was now curling up like a U, and Vincent dropped it into the ashtray. He then, slowly, as if lifting the world, raised his immense girth up so that again his face was in shadow.

"Fuck it," he said. And then he turned his back, walking slowly out through the bar area, and out into the gathering mist.

Epilogue

While Vincent's dramatic destruction of his beloved application made him a legend among CRT's demoralized diaspora, the truth was that he had torched a disk containing nothing more vital than his last résumé. Throughout 2001, Vincent fought a paper war with Aether. He racked up huge legal bills in the process and six months later, found himself no closer to justice. Although his case had merit, Aether filed a flurry of motions that saw to it that any satisfaction that Vincent might obtain would probably occur sometime in the twenty-second century.

In the spring, impatient with his attorney, Vincent took his appeal directly to the people, by distributing a sworn statement alleging colossal malfeasance, cruelty, and theft on the part of Aether. Together with several other people who'd been screwed over, he mailed this document to every member of Aether's board, as well as the company's main shareholders.

But nothing came of it—it was either intercepted or simply lost in the shuffle. In the meantime, Aether has rejected all his requests to settle, and continues to threaten prosecution for theft of services, breach of contact, and libel. As of this writing, Vincent's case remains unresolved. However, to Vincent's consolation, CRT's executive staff was the first to be fired when Aether closed down CRT's offices in August of 2001.

Neither time nor distance, nor Aether's ever-worsening situation (its stock price is now trading at least 95 percent below its peak) have done anything to quell Vincent's desire for revenge. Currently employed as an honest and straightforward developer at a small tech company in Long Island, New York, he claims to be biding his time until the opportunity presents itself for him to clear his name and regain control of his product.

— Shape-Shifters: See Them Change —

Shape-Shifters: Who Are They?

Remember that friend of yours who suddenly went from a regular "9:00 to 5:00 guy" to a caffeine-crazed disciple of all things dot.com? Or that nice, stable woman in the next cube who announced out of the blue one day that she was going to quit her job to become a full-time day trader? Or, most disturbing of all, that person in the mirror—once bright-eyed, smooth-skinned, and wrinkle-free—who now, after doing three hard years on the Internet, could pass for Steve Ballmer's older brother? Well, as much as you don't want to hear it, and as pissed off as we're probably going to make you, all the people we mentioned were Shape-Shifters, victims of a sudden and completely unexpected metamorphosis that changed them from ordinary folks into fanatical Internet zealots willing to sacrifice everything in their lives, including friends, family, mental and physical health, to the Gods of Internet Success. Unlike dot.com workers in their twenties, many of whom had never held a real job before, Shape-Shifters were older folks in their thirties and forties who, when the Net reached critical mass as a pop cultural phenomenon, had topped out in their careers and viewed the New Economy as a way for them to rejuvenate their mundane lives, and perhaps even step off the corporate treadmill forever.

Unrealistic expectations? In hindsight, you bet they were. But at the time anything seemed possible, and Shape-Shifters plunged into the IPO Dream with a vengeance. They studied O'Reilly books, went to seminars, shaved their heads, learned all the buzzwords, and dropped all the right names. They switched jobs, relocated their families, and moved to Swinging Silicon Valley—physically or in their own minds. They took up residence by the thousands in numerous, ill-conceived start-ups intent on becoming the Microsoft of the eGardening space.

All throughout the boom they slapped each other on the back, congratulated each other about "escaping the rat race," calculated their net worth, and put their trust in the Bubble's never-ending expansion. They spent money like it was going out of style, often on elaborate marketing plans that would have taxed the budgets of most South American countries. But, when the Venture Capital Viagra ran out, they hit a wall—a hard one—at Internet speed. The story of Shape-Shifters is not a happy one. Many who gloated about parlaying $50,000 middle management jobs into fast-lane dot.com jobs kicked themselves a few months later, especially after the NASDAQ crash of April 2000, when thousands of them lost their high-paying jobs, inflated titles, and expectations of a plush early retirement. Some crawled back to their old corporate jobs, never to utter the dreaded word "change agent" again. Others simply washed out, leaving broken families, neglected children, and a welter of antidepression drug prescriptions in their wake. The best-adjusted of them went on adapting as best they could to a post-NASDAQ world of radically-diminished expectations, playing air guitar to their favorite sixties pop music, especially the lyrics to "You Can't Always Get What You Want."

Are You a Shape-Shifter?

You might be a Shape-Shifter if . . .

- In the sixties, you were a hippie; in the seventies, you were a disco bunny; in the eighties, you were a yuppie; and in the nineties . . . well, you know the rest.

- Your model of mid-career Internet pole-vaulting was none other than Lou Dobbs, who left CNN to head up Space.com—a nebulous extra-terrestrial Web portal swallowed up in the black hole of 2001. (Dobbs, unlike many making similarly precipitous leaps into cyberspace, teleported back to CNN before his oxygen ran out.)

- The number of jobs you've had in the last two years exceeds the number of jobs you had in your first thirty-five years of life.

- If you could do it all over again, you would chase young girls in a red sports car instead of opting for an early breakout clause in your stock options.

- You used to teleconference every morning with five different brand managers in four different time zones. These days, you'd be lucky to make one phone call a week to the Department of Labor and/or the local liquor store.

- You've given up your subscriptions to *Fast Company*, *The Industry Standard*, and *Wired* in order to keep yourself in fresh toilet paper, Ramen noodles, and budget pornography.

- You thought that everyone at your old job who wasn't interested in the New Economy was a fool—until you began calling for your old job back.

- Since crapping out of the New Economy, you have either become a forty-something version of a twenty-something slacker, or you have reverted to the tight-assed corporate drone you used to be, disavowing any knowledge of your prior, wild and woolly Internet ways.

Fun Facts About Shape-Shifters

Favorite Singer: Madonna (You know, that whole no-talent, constantly "reinventing herself" jive.)

Former Pet Phrase: "Perception is everything." (Translation: "It's not what you know, it's what people *think* you know.")

Current Motto: "That which does not kill you makes you stronger." (Translation: "Goddammit, I didn't get rich, but that's okay, because I'll just hop on the next bandwagon that comes down the pike. By the way, where can I find out more about this whole biotech thing?")

Post-Bust Stress Rating (PBSR): 3.2 (Shape-Shifters would have you believe that "they got soooo screwed." In reality, the thing they hurt most was their pride.)

Strengths: Hard-working, inwardly directed, self-managed.

Weaknesses: Gullibility, unquestioning faith in the mainstream media, a lemming-like devotion to fads.

Most Common Jobs They Held During the Boom: CFO, CTO, Chief Marketing Officer, Editorial Director, Global Head of Sales. (Sounds impressive, right? Well, let's not forget that at your typical struggling dot.com, the CMO, say, was mostly likely clearing rotted Chinese food out of the communal fridge, when he wasn't working on the company's global branding strategy.)

Psychological Profile: Self-proclaimed individualists who were really just going with the flow, grabbing for the brass ring, and thumbing their noses at the traditional corporate world where they were second-rate. (Conversely, for CEOs like Lou Dobbs, it was all about going from rich to ridiculously rich.)

Shape-Shifters: The Story of Gene

Gene cradled a beer in one hand and the remote in the other, waiting for the last fifteen minutes of *The Good, the Bad and the Ugly* to commence. Since the advent of cable, Gene had seen this movie about thirty-five times and, for him, it was right up there with *The Dirty Dozen* as one of his all-time favorite guy flicks.

Gene usually channel surfed during commercials, but pure laziness got the best of him this particular evening and as the colorful parade of brand messages unfurled, the screen on his fifty-three-inch Sony Projection TV suddenly went black. He was about to get up and check the cable connection when a deep, ominous, monotone voice boomed from his Bose speakers, almost knocking him out of his seat.

"This is your future. A world without hope, without light. A world where there is no Social Security, no health care . . . "

The voice continued for several more seconds, waxing more and more Apocalyptic, now over a black-and-white background of people writhing in pain in a fiery pit. Then, when it seemed like the negativity would go on forever, the voice stopped midsentence and the montage of hellish images was replaced with the logo of an online broker whose tagline read, "Take back your future."

It lasted less than a minute, but when it was over, Gene was in a sweat. Laura, his wife, who had a habit of hovering around him when he was watching TV, immediately knew something was wrong.

"Are you all right?"

"Yes, I'm fine."

"You don't look so good."

"I think I ate something I shouldn't have."

"I told you to stay away from the stuffed peppers."

"Yes," Gene said weakly, aware that something major had just happened in his life, although he wasn't quite sure what it was.

Getting Whacked

The next day at work Gene was a complete mess. He had been up the entire night tossing and turning and was so bleary-eyed from exhaustion that several of his coworkers wondered if he was coming down with the flu.

"I'm okay, really. Thanks."

"Are you sure? You're really pale. Maybe you should go home."

Home was the last place Gene wanted to be. If he were home, chances are he'd get bored and turn on the TV and end up watching that horrible commercial again. No, he'd load up on caffeine and make it through the rest of the day the best he could, secure in the knowledge that, since he was so exhausted, he'd be asleep tonight as soon as his head hit the pillow.

Unfortunately for Gene, things didn't work out that way. He crawled into bed right after dinner and, instead of dropping off, he found himself tossing and turning again, growing increasingly unhinged because the harder he tried to fall asleep the more awake he became.

"What's wrong with you?" Laura asked, finding him in a pool of sweat a few hours later.

"Don't you see? I'm a dinosaur!" Gene almost never raised his voice, and instantly felt bad about having done so. Sitting up in bed, he tried to calmly explain how the commercial from the night before had made him realize that he was being left behind by the New Economy. Sure, he was doing well, and had worked his way up the ranks in Procter & Gamble to Senior Brand Manager. But despite his impressive title and fat salary, he still felt like a failure. He was spending his entire life overseeing corn chips and scented tissue, while other people were building the future online, and getting very rich doing it.

"That's complete nonsense, honey," Laura said, running her hand through his hair. "You've been a great father and husband."

"But what if we don't have the money to send the girls to a good college? What if we don't have enough money to support ourselves when we get old? What then?"

"You've got a good, stable job. If we don't have enough money, we can always take out a second mortgage."

"I could've done better. I can do better. I . . . "

"Gene, do you realize what the problem is? You're having a midlife crisis!"

"No, I'm not. If I were, I'd be chasing young girls in a red sports car."

"Fine," Laura huffed, turning violently away from him in bed. "Make jokes if you want. But have you considered confronting your fears, instead of running away from them?"

Take 5.0

Against his better judgment, Gene let Laura convince him to go to a party in nearby Indian Hill a few nights later. "It'll be good for you. You'll forget about your troubles," she promised him.

Gene hated parties, especially ones in Indian Hill, the Beverly Hills of Cincinnati, where the top execs at Oscar Meyer and P&G lived, and where he was certain to run into someone who'd annoy the shit out of him. Gene had been to numerous soirees in this tony burg before and they were all more or less the same, with the guys blabbering on and on about their golf game and Cuban cigars, and the women waxing Martha Stewartesque, extolling the virtues of homemade flower baskets and churned butter.

Gene wasted no time preparing himself for the worst. After greeting the hostess and handing her his coat, he made a beeline for the bar, which had been set up in the corner of the living room.

"Give me five shots of Jack Daniels."

"Pardon?" the tuxedo-shirt-and-bow-tie-clad bartender asked.

"You heard me," Gene commanded. "I'm gonna need it."

He downed the shots one right after the other, then had the bartender pour him a glass of Chardonnay as a cover.

Thirty minutes later, when his buzz was just kicking in, the room had filled up with roughly forty people; he knew about half of them from the office, and the rest were either friends or associates of Laura's from her part-time job at a local law firm. Although Gene had lived in the suburbs for most of his adult life, he still felt like the poor kid from South Philly who had crashed the Great American Middle Class Party. He compensated for his insecurity by being overly friendly or, as in this situation, drinking heavily in order to maintain the facade.

"So how are you doing this evening?" he asked a tall blond-haired guy, who looked to be his age, and whom he had seen in the company cafeteria a few times.

"Not bad," the guy responded. "Say, are you in the New Media Division?"

"No, Brand Management, Consumer Paper Products, Scented Household Goods."

"Have you guys come up with an Internet strategy yet?"

Gene froze for a moment. He now realized that he was probably talking to some senior-level person and needed to come up with a thoughtful answer.

"My understanding is that we have been waiting for the medium to mature before doing anything that would damage our brand equity."

"I think we need to revisit this issue," the blond-haired guy said curtly, taking a bite of the cube of cheese he was holding and disappearing into the crowd without another word.

Gene let out a sigh of relief. His years of back-to-the-wall bullshitting, combined with the booze, had once again saved his ass. Or so he thought. As it turned out, however, he spent the rest of the evening running away from innumerable conversations about the Internet: in the kitchen, in the living room, on his way out of the bathroom—everyone everywhere at the party seemed obsessed with how rich the New Economy was going to make them, and for Gene there was no escape.

"Do you think Amazon is really going to hit 400? Should I sell now or wait?" asked a forty-something woman with store-bought breasts.

"Juniper is really leaving the old-line networking companies in the dust," said a young man inexplicably clutching a copy of the *Wall Street Journal.*

"My nephew's company just filed for its IPO," boasted a silver-haired gentleman. "With all the stock options he's got, he should be a millionaire in six months."

As drunk as he was, Gene tried repeatedly to steer the conversations toward the more familiar topics of golf, cigars, and home improvement. To his amazement, such Indian Hill staples no longer seemed to matter. Gene came very close to telling several of these people to lighten up and have a drink, but wisely held in his frustrations until the ride home.

"Did you see them?!" he raged from the passenger seat of their Volvo wagon. "They all had the same maniacal glint in their eyes."

"What's it to you?" Laura said, tightening her grip on the steering wheel.

"I've been hearing this shit nonstop for the past year. I'm sick of it and sick of being reminded of what a loser I am."

"Oh, here we go again. You and that damn commercial."

"It's not the commercial. The commercial was just the last straw."

"And what was tonight then?"

"The beginning of my revenge," he said, sounding a lot like Clint Eastwood.

Like Tony

Come Monday morning, Gene marched into his boss's office and gave him the skinny on his brief conversation with the blond-haired exec over the weekend. His boss, who usually didn't get caught up in corporate gossip, immediately pulled up the marketing org chart on his computer and began trying to figure out who it was Gene had talked to. After about fifteen minutes of going though the layers upon layers of management comprising the P&G marketing machine, his boss narrowed it down to three people, all of whom were very senior-level.

"So what should I do?" Gene asked, already knowing the answer.

"We really need to put together a proposal. I know you're not exactly an Internet expert, but do you think you can handle it?"

"No problem," Gene said, and spent the next week devouring everything he could get his hands on, from Web sites to tech magazines, manuals, books, even software he barely knew how to use. When he was done, he presented his boss with a one hundred-page document that included a comparative analysis of what other companies in the consumer goods business were doing, demographic breakdowns of Web usage, the hottest areas in the industry, as well as the general perception of traditional brands online.

It was quite a piece of work, especially from someone who until just seven days before didn't know the difference between a browser and an e-mail client. Gene's boss praised his efforts, and Gene felt very, very good about himself. In a week, he'd become more knowledgeable about the Internet than anyone else in his department, perhaps the entire company.

"They're going to give you a big promotion," said Laura, scooping extra mashed potatoes onto his dinner plate.

"If anybody gets a promotion, it'll be my boss, and then the report will get buried for the next year."

Sure enough, that's exactly what happened. Within a month of submitting the report, with his name in bold letters at the top, Gene's manager got kicked way upstairs, never to be heard from again. Although several people in his department came by to console Gene and tell him that it should've been him getting the promotion, Gene was very nonchalant about the whole thing.

"It's all right. You've got to make your own plans," he said with a poise and a confidence none of them had ever seen in him before, causing many to wonder what had come over their normally demure colleague. In the months that followed, there would be more changes, some quite radical, which confused the hell out of everyone, particularly Laura.

"You're on the computer all hours of the day and night. You've subscribed to every technology magazine on the planet. You've even been going to the gym and you've bought a closetful of those stupid blue Oxford shirts. What's going on, Gene? Have you flipped your lid?"

"It's part of my OS upgrade," he said.

"Would you mind repeating that in English?"

"I'm facing my fears, just like you told me. Don't worry. I'll make you proud."

She tried pressing him further, but he would say nothing more than "I'm going to have a big surprise for you very soon."

Getting Silly

The "surprise" came about a month later. Gene snuck up behind Laura while she was working in the garden and began kissing her neck.

"Did I ever tell you that you're as beautiful now as the day I met you?" he cooed.

"Stop, silly. The neighbors will see you."

"Let them. We'll put on our own porno movie out here."

"Gene, are you drunk?"

"No, even better. I did it."

"Did what?"

"I've landed interviews with three of the hottest Silicon Valley start-ups. I'm flying out tomorrow morning."

"Do you think you could've waited a bit longer to tell me that you're leaving, to say nothing of the fact that you're contemplating a major career change?"

"I'm sorry. I really didn't know until an hour ago. That's how Internet Time is, you know."

"Okay. Go for it. But don't accept anything without checking with me first, understood?"

"We're going to be rich, rich!" he screamed before skipping back into the house to start packing for the trip.

By the next night Gene was on the phone with Laura, boasting that he had already received four offers.

"Four? I thought you were interviewing with three companies."

"Yeah, but on the plane I started talking to a guy in the next seat and it led to an interview with his CEO at Toshi's Sushiya—Larry Ellison's favorite sushi restaurant."

Gene then proceeded to run down the details of the compensation packages and Laura could barely believe her ears.

"So the pets e-tailer is offering $200,000, with 300,000 options at a twenty-five-cent strike price; and the infrastructure company wants to give me a BMW roadster as a signing bonus in addition to . . . "

Gene went on and on, becoming more and more excited, and soon the excitement had rubbed off on Laura, and they were both screaming like two lunatics who had just won the lottery. Before getting completely out of control, Laura's legal instincts kicked in and she made him promise not to sign anything without first letting her review the documents. Gene agreed, and less than an hour later, the fax machine at their home was working overtime to process the pages upon pages of employment agreements, stock options statements, and relocation packages. In a hint of what was to come, Gene had scrawled, "We need to make a decision by tomorrow" on the cover sheet. Laura wasn't happy over such an insane deadline but, with thoughts of what they could do with all that money, she accommodated and, before long, was calling Gene with her revisions and suggestions about which was the best deal.

Getting Out

"So, you like the pets e-tailer, but you think they need to be clearer on the breakup language? Okay, baby. I'll make sure they clear that up," Gene said, the reception on his cell phone at the point of breaking up.

"How do you want to handle your resignation from P&G?" Laura asked amidst the static.

"Don't worry about it. I'll send them an e-mail."

A second later he was gone. He made it seem like the connection had failed but, in truth, he didn't want to talk to Laura anymore because he was lying through his teeth. He had executed the agreement with the pets e-tailer and tendered his resignation hours before for fear that any hesitation on his part would be an excuse for them to hire one of the other candidates. To say that the competition for the senior-level, $200,000 job was intense would be a massive understatement. Gene was interviewed, screened, and prodded by everyone from the CEO to the CFO, the CTO, the Board, even the VC, and the mood was that the other candidates would've sold their mother to land a CMO slot with a hot, prepublic start-up. It took him three marathon days to get through the tunnel, but success was at the end of the tunnel—he nailed the job.

Starting from Zero

Gene did more work in the next few weeks than in his previous twelve years with P&G. As employee number six—and the sole member of the Marketing Department—he was literally starting from zero with two projects to balance: working a design firm to come up with a corporate identity and logo, and drafting a marketing plan that would redo the "About" section of the new Web site. He had already reserved booth space for several upcoming trade shows and short-listed three PR agencies to handle media.

What made it tough from the get-go was a people shortage. Although Gene had been given approval by the CEO to hire five assistants, he had not gotten around to doing so because he had his hands full with everything else, including relocating his family to the corporate apartment in Foster City, where he had been living since his second week there. And organizing the move had been a nightmare. Laura, who was annoyed by anything that was out of place, was flipping out.

"Are we moving or not?" she called in, echoing around the office before Gene could route it from the speakerphone. "Our bags have been packed for the past month!"

"We're moving, we're moving," said Gene, fixing his earpiece to his head. "I just have to finish nailing down the specifics with the relocation company. It won't be more than another week, I promise."

One week rolled into the next and it was a month later when the moving guys pulled up in front of the corporate apartment to unload the few possessions that would fit into its tiny confines. Gene knew that there would be hell to pay when Laura arrived with the girls and saw how small the place was, but he was so tired from going without sleep for the past four months that he couldn't worry too much about Laura's anger until it was right up in his face. There were five thousand more pressing issues to take care of, such as the stench rising from the pair of pants he had been wearing for the past week straight.

"I moved out of a five-bedroom house for this?" she asked, taking one look at the meager accommodations.

"It's only temporary until we find a regular apartment," Gene stammered.

"You said it was a big two-bedroom. This is a small one-bedroom!"

"We can use this as a bedroom," he said, pointing to a small space next to the kitchen.

"Are you crazy? When was the last time you did the dishes?"

"Well, I . . . "

"We can't sleep here. We're going to a hotel. Where are our clothes?

"In the bedroom."

Gene, for all the excitement, felt like he was about to fall asleep standing up. And he probably would have, if Laura hadn't completely lost her mind when she went into the bedroom and discovered that most of their belongings were in storage.

"Call us a taxi. We're going to the nearest hotel."

"That won't be necessary. You can use the Mercedes SUV the company leased for us," Gene said, hoping that this little token of corporate beneficence would make up for the sorry state of their living arrangements.

She stormed out, just as Gene's cell phone rang.

"No, I'm fine," he said into the shell-shaped device. "I was just getting my family settled in the apartment. No, I'm not busy. I'll be right there. In the meantime, tell the designer that we need to work up a treatment for . . . "

Web Widow

Gene didn't speak to Laura for the next several days. He called her cell phone repeatedly and left countless apologetic messages, but she refused to pick up or call him back, still steamed over the whole apartment situation. When they did finally talk, she was civil and he made sure to grovel. If Gene had been in his right mind, he would've realized that he was well on his way to ruining his marriage. But, like many people who had been bitten by the Internet IPO Bug, Gene was far from being in his right mind, and would only realize how truly warped he had become when it was too late. "I'm really, really sorry, baby. Don't be angry. I'm doing all this for you and the girls," he said, sounding very penitent.

"All right," she replied. "Just make sure it doesn't happen again."

"It won't. Now where are you? I'd like to give you a little present to make up for all the trouble I caused." He scrawled the address of the hotel into his Palm and surprised her that night with a brand-new Audi wagon and a stack of real estate listings.

"Go out and rent the biggest house you can find. Money is no object."

Gene's generosity smoothed things over considerably, but it wasn't before long that Laura was furious at him again. He was working constantly and saw his family only long enough to grunt and stagger off to bed. Laura tried confronting him on their lack of "quality time" and found it impossible to talk to him when he was so tired—during the day was equally frustrating because he was so busy. On the rare occasions that she was lucky enough to reach him by phone, she could barely utter three words before he had to rush off to a meeting or attend to some "mission-critical" issue. Short of showing up at his job and having a massive fight in front of his coworkers, she was reduced to venting her frustrations via e-mail:

> Gene,
> Was that you I heard coming in the door last night at 3:00 A.M.? Or was it a burglar?
> Love,
> Your Web Widow

She expected a call. Or even a response. Instead, her mail bounced right back to her with the following error message: "Could not reach recipient. Recipient's mailbox is either full or is an unknown user. Will continue trying for the next twenty-four hours."

Get Broke Fast

To be fair, Gene really was up to his neck. In the brief eight months that he had been working for the pets e-tailer, the company had gone from ten employees to well over one hundred, and had just filed to go public. With the CEO, the bankers, and the board breathing down his neck, he would have to generate as much hype as possible in the next six months. This was no small task, considering that every other dot.com out there was going to outrageous lengths to get noticed or, as they put it, "build brand equity." Faced with such insane competition for eyeballs, Gene knew that he had to pull out all the stops, or fail miserably, so he wasted no time arranging meetings with the most prestigious agencies in New York City to plan a massive ad campaign. If the campaign was a bust, and didn't generate enough hype, they would be out on their asses, and the personal and professional sacrifices they had made for their precious stock options would be for naught. But it also was a win-win situation. If the campaign turned out to be a fabulous success, he'd be stinking rich. And if it failed, well, then at least he'd had his big shot to bring a brand he had created to millions of people around the world—something he would never have achieved if he had been back at P&G, shepherding someone else's brand.

Gene's whirlwind of meetings up and down Madison Avenue took place in mid-1999. Although Christmas was months away, they had to get the campaign together now and buy the appropriate media slots for the most crucial shopping season for retailers, electronic and otherwise.

Trouble on Mad Ave

The first sign of trouble in Gene's new career soon made its appearance. Everyone hated the mockup from the agency Gene favored. The work featured a dog with sunglasses sitting on a beach, which those assembled either found "too tame" or reminded them "too much of the Lycos dog." Gene took the criticism in stride, but when they suggested that they go with a Silicon Valley–based agency with a longer track record with tech companies, he was livid. Hadn't they hired him to handle the marketing? Wasn't he the one with the consumer experience? Since when did the ability to hack code make you a branding expert?

Although a tiny voice inside his head told him to read these people the riot act, his underlying insecurities, his greed, and his unwillingness to possibly walk away from his dream job worked against his common sense, and he was soon dialing the lead person at the agency everyone favored, hoping to get the campaign rolling.

The committee approach, which had overruled his agency choices, only intensified as the campaign unfolded. Either the CEO didn't like the message, or the VC hated the typography, or the CTO missed the point entirely. Gene was forever doubling back on himself until it got to the point where he felt like he was completely ruled out of the process and the company would've been better off letting the CTO draw up the campaign in PowerPoint.

Dog Days

Things should have improved with Laura. He now had more time—and plenty of trappings, including a big, grand house and two private schools for his kids—to rekindle their love. But they fought more now than ever.

"What do they want from me?" he raged, holding his fifth beer. "What do they want?!"

"Gene, this is what you wanted," she snapped, provoking an extended verbal firefight that would drive him from the room. A few days later, just when he thought things couldn't get much worse, Gene was summoned into the conference room to review the final treatments for the TV, print, and billboard campaign. As the CEO, CTO, and VC rhapsodized over how "brilliant" the whole thing was, Gene couldn't help but feel sick. The print and billboard collateral consisted of a photo of twenty dogs standing in front of a large hydrant emblazoned with the company's logo; and, what was even more horrible, the commercial was shot cinema-style, with shaky, hand-held cameras, and featured the same dogs from the print and billboard campaigns breaking the chains of their masters and running madly to congregate at the massive fire hydrant. It was amateurish and stupid. Worse, it raised the issue of why they would subliminally suggest that customers would want to lift their hind legs and piss on their logo!

"So they're coming to the site and leaving?" Gene asked the senior design director.

"Yes, most of them seem to be stopping at the home page before clicking away."

Gene relayed this information to the CEO and his response was to tell the CTO to have someone improve the load time. This seemed like a sensible move to Gene, who knew that it was really too late to do much else. The immediate result was that sales did improve, but not by much, not enough to convince Gene that the campaign was going to be anything other than a disaster. "Are we doomed, Gene?" asked one of his team, seeing him sitting at his desk looking like he was about to jump out the window.

"Not yet," he responded. "The commercial will be the real test."

Super Sunday

The night before the Super Bowl, Gene was up tossing and turning as furiously as he had 6,000 Internet years before, after he had seen the spot for the online broker. Laura, although her anger at him had by no means abated, took pity on him and got up and made him a glass of warm milk.

"Here, drink this," she said, offering him the glass.

"Is there any Kahlúa in it?"

"Just drink it. I'm not in the mood for an argument right now."

Gene guzzled the milk like a man dying of thirst. To his pleasant surprise, he found himself getting very drowsy afterwards and falling off into a deep, peaceful sleep. Awaking the next morning, he felt better than he had in ages—almost good enough to prepare himself for the spectacle that would unfold at approximately 8:42 that evening.

Gene clutched a throw pillow as the first frames of the commercial flashed across the screen. "Hey, come look, it's Daddy's commercial," yelled one of his daughters. Laura reluctantly joined them in the living room.

"Daddy, are the dogs going to do pee-pee?" asked his other daughter, the youngest.

"Just watch," Gene said. "You'll see."

When it was over, both daughters were disappointed that the pooches hadn't relieved themselves, and so was he, although he was soon saying otherwise into his cell phone.

"Did I see it?! Of course I saw it! It was great!"

The phone rang incessantly for the next hour and Gene shouted his feigned enthusiasm to the CEO, CTO, the VC, and several colleagues before finally sharing his true feelings with Laura.

"That was the worst piece of crap I've ever seen."

"C'mon, Gene. It was no worse than the other commercials."

"That's the problem. It was just as bad as the rest because they were all cranked out by the same few agencies that have been cleaning up on dot.coms."

Gene went on to explain that the commercial should've explained how fun and easy it was to buy pet supplies online, instead of wasting such exposure on mere branding.

"I'd be surprised if we made one sale."

"You're just being crazy."

"I hope so. I really hope so."

Gene would soon learn that his estimate wasn't too far off. The commercial, which had cost them in excess of $10 million, had tripled traffic, but only generated 1,000 sales in the first week after its airing. Gene knew what was coming, and rather than nervously wait it out, he confronted the CEO late one Friday as he was preparing to go home.

"We're in deep doo-doo," said the CEO. "Maybe if you had listened to me the campaign would've been more successful."

"I don't know what you're talking about. This was all your idea."

Dead Meat

Gene stopped off at a bar on his way home. Numerous beers and several hours later, he stumbled through the door, carrying a box of belongings, and didn't get as far as "hello," when Laura figured out what had happened.

"You lost your job, didn't you?"

"Well, I wouldn't exactly say, 'lost'—it's not like I misplaced it or anything."

"You've been drinking."

"Yes, I have."

"I'm beginning to think that you're an alcoholic, Gene."

"Why don't we discuss it over a beer?" he replied, jokingly.

"Oh, you're a regular comedian," Laura shot back. "Tell me something, all bullshit aside, did they or did they not give you a severance package?"

Gene stared at the rug for several seconds before responding very slowly. "Yes, they did, but it's minimal—one month's salary."

"A month?! What happened to the breakup language I added to your employment contract?"

"Honey, I have a confession to make. I never included that language. I signed the contract right away for fear that I'd miss out on the position."

A few days later, there was a knock on the door of the motel room where he'd gone after Laura had thrown him out. Although he hoped it was the maid, he had a pretty clear notion of who it really was. Cracking the door slightly, he came face to face with a burly Mexican dude with a gun.

"Mr. Kraft, I'm a process server . . ."

Epilogue

So ended Gene's marriage. Considering how things had gone since they'd moved out to California, he wasn't entirely surprised, although he'd hoped that they could've worked things out. In the months that followed, Laura and the girls moved back to the old house in Cincinnati and Gene, who decided to settled the divorce quickly through mediation, moved into a small, one-bedroom apartment in San Mateo, completely clueless about what he was going to do next.

Gene hit the skids for a while. Just as he was beginning to pull himself together, and think about getting another job, the NASDAQ crashed, putting an end to the New Economy as he and everyone else had known it. In place of interviews, Gene found himself hanging out at pink-slip parties in San Francisco with people who seemed even more demented than he was. Realizing that, if he didn't cut back on his drinking, he'd end up in the gutter, Gene decided to dry out and sign up for Recession Camp, an outdoor getaway whose purpose is to help unemployed dot.commers stay "healthy, happy, and sociable." After going on numerous hikes and attempting to improve his diet, Gene slowly settled into Northern California's odd, "we don't care where you came from as long as you're cool here" life ethos. Attending AA meetings and hanging out with weirdo former dot.commers also helped and led him inescapably to Buddhism. There was a temple right in San Mateo where he lived, and he made sure to visit it every day, even if he wasn't in the mood to chant. There, he met Jessica—a redhead. She was a designer and she was always turning ordinary objects into art, without a second thought about monetary gain. In this respect she was the opposite of Gene, although she was always receptive to his ideas.

One night, finishing up dinner at Jessica's, he stumbled across something she had made as a lark and he immediately had a brainstorm. It was an antique white phone to which Jessica had attached pink fur.

"This is brilliant," he said, with a laugh. "Have you ever thought of selling it?"

"Uh—no."

So came the idea for "FurryFones"—$120 stuffed animal-like phones that Gene and Jessica now sell to local independent bookstores and souvenir shops. They make a pretty penny for their efforts, enough for Gene to keep current—more or less—with his child support payments. If anyone had told Gene just a few months earlier that he'd be earning his living gluing fur to phones, he would have said they were crazy. The more he thought about it, though, the more he realized that it made a hell of a lot more sense than selling dog food online.

5

Aliens: Twice the Work for Half the Pay

Aliens: Who Are They?

These techies arrived in America in the late 1990s to find fortune in software and, while some found the American Dream, others found a peculiar nightmare.

You've doubtless read many success stories about how e-enabled immigrants struck it rich in the United States. For instance, there's the rags-to-riches tale of Sabir Bhatia, the founder of Hotmail, who in a few short years had transformed his vision for a free e-mail service into a $200 million check from Microsoft. Or Venky Harinarayan, who sold Junglee, an e-commerce software company, to Amazon for $180 million.

And let us not forget Pierre Omidyar—the founder and chairman of eBay—who is now one of the richest men in the world under forty, according to *Fortune*.

Sure, many immigrant techies got rich, but for every Pierre Omidyar or Sabir Bhatia, there were tens of thousands of foreign nationals laboring in the lowest ranks of the tech economy with no hope of upward mobility. On the surface of things, these workers were doing what many of our grandparents and great grandparents had done: pursuing the American Dream during one of the most dynamic and exciting periods in our history. In reality, these workers—known as H-1Bs (according to the class of the guest visa they held)—were the living and breathing property of an evil breed of human importer-exporters known as "body shops." These double-dealing middlemen lured thousands of ambitious tech workers to the United States with ads promising exciting opportunities that often didn't exist.

When the New Economy began to crash in the spring of 2000, H-1Bs were the first to be laid off. They weren't offered severance packages, or a free laptop, or use of the company's facilities to look for another job; they were simply cut loose, which put them in imminent and real danger of being deported. According to U.S. immigration law, guest workers who lose their jobs only have ten days to find a new one, or be forced to leave the country. The result was that many pink-slipped H-1Bs had to pack their bags and give up all vestiges of the lives they were trying to build in the United States. Even those who were lucky enough to find another company to sponsor their stay ended up getting screwed, because switching jobs reset their naturalization time clock to zero.

Yes, many H-1Bs had positive experiences and were able to put down roots in the United States, or return to their native lands with extraordinary nest eggs. But many endured working and living conditions that were throwbacks to the late-nineteenth-century sweatshop era, with five or more crammed into fetid, warrenlike apartments. They also were routinely browbeaten and abused by employers, who knew that they couldn't fight back, or speak out, because the slightest hint of dissent could land them on the next plane home. And even their coworkers often treated them like dirt, either because they had dark skin or because their very presence implied a message from management that said, "We're two steps away from giving your job to someone who works hard, never asks for a raise, and would never utter the word 'Union.'"

Although most foreign nationals no longer consider the streets of Silicon Valley to be paved with gold, globe-trotting tech workers continue to seek their fortunes in any land where a modem and a paycheck might be found.

Are You an Alien?

You might be an Alien if . . .

- Your duties at your last job included synchronizing data sets, installing the latest Apache mod, and washing your boss's Boxster.

- Someone promised you that you'd be making love with Drew Barrymore in the hot tub of your limousine by the second quarter of 2000.

- Your company's pink slips are printed in seven different languages.

- You're on the books at $100 an hour but your take-home pay amounts to less than the weekly allowance of a high school kid in Palo Alto.

- You're so bugged by the fact that your coworkers are incapable of pronouncing your name that you prefer to be called by your IP address.

- You left your native land to escape civil unrest, power blackouts, and political corruption. (The good news is that you feel right at home in California.)

- You have delightedly discovered that American women are more than twice as intelligent as the goats in your country.

- You were hoping that your American buddies would be cool-guy babe magnets like George Clooney and Brad Pitt, but have discovered that most look like a cross between the Stay-Puft Marshmallow Man and Homer Simpson.

- You hate going home at night, not because you have six roommates sharing a one-bedroom apartment, but because you can't watch the news without getting pulled into a heated argument about global politics that can easily devolve into an international food fight.

Fun Facts About Aliens

How They View Themselves: Over-achieving idiots who let themselves be fooled by smooth-talking, amoral recruiters.

How Other People View Them: Coolie labor, temporary help, disposable human capital, and now (in the wake of September 11), people who need to be watched very carefully.

Post-Bust Stress Rating (PBSR): 10 (which, if you recall, is the number of days Aliens have to get a new job, or get on a plane before the INS comes calling).

Mode of Dress: Normal—in fact, super-normal. Aliens, conscious of the fact that they're strangers in a strange land, overcompensate by buying all their outfits right off the rack from Today's Man.

Favorite Things about America: The Grand Canyon, the Mississippi River, unlimited access to pornography, and Regis Philbin.

Least Favorite Things about America: Bland food, outspoken women, and bad beer.

What They Do for Fun: Make transoceanic phone calls, fantasize about the hardwood floors in the house they'll someday buy, and engage in hand-to-hand combat with the bacteria in the back of their shared refrigerator.

What They Did During the Boom: Work, work, work, work.

What They Did After the Bust: Pack, pack, pack, pack.

Where They Are Now: Whether they are back in their home countries licking their wounds or off pursuing another dream job in another foreign land, they are happy to have survived ordeals that would've killed your average foosball-playing, Snapple-guzzling, free pizza-loving American techie slacker.

Aliens: The Story of Dana

Dana was feeling antsy. She knew that the Australian company she worked for was undervaluing her skills, and she also knew that better, higher-paying tech jobs were plentiful in the United States. So in July of 1999, she signed up with a recruitment agency that specialized in bringing foreign technology workers to the United States.

"It's a dream job for somebody like you," said Phil—a guy in his thirties. "You've got three years of Sybase; two years of Oracle. I don't think I've ever seen a more perfect match."

"Where will I be working?"

"Right in L.A. Orange County," said Ray—a middle-aged fellow in a suit. "It's a three-month assignment—a top-notch company, an excellent pay rate. Plus you get your own car, your own apartment, the works."

"It sounds great," said Dana.

"Your skills are stellar," said Phil.

"Topflight," said Ray.

"Thanks, guys," said Dana. "There's just one thing."

"Yes?"

"This job. It's not clear to me from the description whether it's right up my alley. I mean, the pay is great and the perks are terrific. I just want to make sure I'll be working in database-to-Web development. That's where my real career track is."

"Absolutely," said Phil. "This company is the largest reseller of Microsoft products in the L.A. basin, and that's their whole strategic direction."

"So where do I sign?" asked Dana.

"Right here," said Phil, pushing a twelve-page contract across the table. She picked up a pen and the die was cast; Dana was coming to America, the land of e-opportunity.

America's Guests

Dana, an Australian, was making a decision similar to that being made all over the world by foreign tech workers hoping to work in America's superheated tech economy. By the late 1990s, the tech industry was expanding so quickly—so many companies were being formed and so many jobs were being created—that industry lobbyists feared that a worker shortage might crimp its untrammeled growth. The ITAA (Information Technology Association of America) pegged the shortfall at some 340,000 positions going unfilled each year. The Hudson Institute, the same think tank that had brought Dr. Herman (Strangelove) Kahn to prominence in the 1960s, predicted direly that the shortfall could cost the New Economy more than $200 billion unless quickly addressed.

Many critics disputed the reality of this shortage, claiming that it was the invention of the greedy New Economy's CEOs, whose motivations were simply to rein in the upwardly creeping salaries of America's indigenous IT professionals. "There is no shortage of tech workers," one anti-immigrant activist proclaimed. "There is only a shortage of cheap tech workers." The critics characterized these guest workers—known colloquially as "H-1Bs" for the work permits they carried that allowed them to work in the United States—as "indentured servants" who, wittingly or not, were stealing jobs from qualified American programmers, depressing their salaries, and writing crappy code to boot. The battle over H-1Bs became a high-stakes issue in 1999, when presidential candidate John McCain made H-1B reform a key pillar of his program, declaring that efforts to limit the number of foreign tech workers allowed to work in America was "lunacy."

The tech industry ultimately won the battle to increase immigration quotas. All over the world, consulates were flooded with tech workers applying for guest worker visas. By 2000, almost a half million of them worked in America, in Silicon Valley, the L.A. basin, and wherever else the Information Superhighway needed building. Microsoft, Intel, Oracle, Motorola, and other first-tier tech companies loved these H-1Bs and hired them by the planeload; they were skilled, worked hard, spoke English well, and didn't complain.

In the Pipeline

Dana spent a few weeks getting her things in order before leaving for America. After giving her employer the customary two weeks' notice, she sold some personal possessions she'd no longer need—among them her car, apartment, and extra furnishings. Now, poised with a rolling luggage cart, she stood on line at the Qantas check-in desk at Sydney airport, waiting for the recruiter to deliver her a packet of up-to-date paperwork: work permit, employer contact information, car rental data, and other travel documents.

The plane was due to leave in thirty-five minutes, which made Dana a bit nervous. But she managed to minimize any last-minute nail-biting by passing time with a guy she'd happened to bump into while waiting in line. Like her, he'd also been recruited by Phil and Ray to work in America.

"So what job are they sending you to do?" asked Dana.

"Database administration," said Paul.

"Well, I hope we don't sit together on the plane," she joked. "We've got absolutely nothing in common." Suddenly, she felt a hand on her shoulder.

"Dana!" said Phil, out of breath. "We've got a small problem," he said, fumbling for two travel packets buried in his attaché case. "Unfortunately, we weren't able to get either of you your final working permits—I don't know the details but there were some backlogs with U.S. Immigration."

"So the trip is off?" asked Dana.

"No, no, no," said Phil. "But we've got to, uh, well we've got to clear this up once you're both on the other side of the pond."

"So how do we clear Customs?"

"Basically, when they give you that little form on the plane, form I-whatever, just fill it out and say you're entering the United States 'on Holiday.' All right?"

"On Holiday?" said Dana. "But that's not on the up and up."

"It's just a formality," said Phil, closing his attaché case. "Oh, there's one more thing."

"Yes?"

"Unfortunately, we weren't able to arrange for you to get your own place—not for now, anyway. So, you two are going to have to double up."

"Double up?"

"Yeah. On both the apartment and the car. We'll fix it as soon as we can," said Phil. "Look, call me when you touch down. I'm very sorry, but I really must run now—the car's double-parked outside and I'm going to get a ticket."

"Ahem," said a woman behind them, motioning to the gap that had opened up in the line ahead. Dana and Paul moved on.

"What do we do?" asked Dana.

"I don't know," said Paul. "I guess we get on the plane."

Gauntlet

The flight between Sydney and Los Angeles was thirteen and a half hours long, providing plenty of time for Dana and Paul to get to know each other.

"I don't know what you're thinking," said Paul. "But you don't need to worry about rooming with me. I'm a gentleman."

"Well I wasn't thinking about that. I was thinking about how we're not even in the fucking country yet, and we've already been screwed."

"It's only for three months."

"It might be a lot shorter than that, if the authorities happen to discover our little evasion. We might get deported in about ten minutes."

"You know, Dana, I think I've discovered a strong negative streak in you."

"Negative? I'm a fucking Sybase person. We happen to care about rules."

"Sshh," said Paul. "Keep your voice down. Besides, I wouldn't worry too much about being picked up at the border. The Yanks don't really have the computer horsepower to check more than a handful of immigrants out. Their whole system is based on paper. Millions of pieces of paper."

"Well, you've brightened my day considerably, Paul. Now if you don't mind, I'm going to see if I can get some sleep."

About an hour before the plane was scheduled to land in Los Angeles, the stewardesses came back and handed out small white cards—labeled "Form I-94"—among the passengers. Dana and Paul filled them out and dutifully wrote "Holiday Purposes" in the designated section.

"I've got a bad feeling about this," Dana said.

When the 747 finally pulled up to the gate, Dana and Paul were jostling with about two hundred other Australians snaking their way through a low-ceilinged room staked out by officials of both U.S. Customs and the Immigration and Naturalization Service. The Customs man—looking quite fearsome in his uniform, replete with epaulets and a peaked cap—looked quickly at Dana's passport, her completed I-94, and handed them back to her in ten seconds.

But the INS guy, a white-shirted guy with a badge who looked almost exactly like Jack Webb, was not so forgiving. He carefully looked at her passport, frowned, and waved it in her face.

"Why wasn't your passport stamped by the Customs Inspector?"

"I thought it was," Dana said.

"Please return to Customs, ma'am. Next in line."

Dana went back to Customs and spoke to the man with the peaked cap.

"You didn't stamp my passport."

"You didn't fill out an address," he said.

"But I don't know my address. A guy's meeting me at the airport, and he's going to be telling me where I'm staying. So how could I know my address?"

"If you can't provide an address, I can't stamp your passport," the Customs guy said. "Sorry, but those are the rules."

Dana was trapped. She couldn't go back to the plane, or go forward past the INS guy. She dimly wondered whether she shouldn't simply turn herself in to the nearest airport cop, confess the whole charade, and hope to be deported without being strip-searched. Paul returned a moment later—he'd been told the same thing.

"What do we do?" whispered Dana.

"I don't know. I'm sweating like a pig. The guy up there at INS—the one who looks like, like . . . "

"Jack Webb," said Dana. "Yeah—I know he suspected something."

"This is no joke," said Paul.

"I know," said Dana.

Suddenly, a Qantas stewardess appeared, rolling a small luggage cart. Seeing a fellow Australian in uniform must have given Dana some measure of comfort, because she ran over to her and quickly explained her plight.

"Just say you're staying at the Los Angeles Airport Days Inn," the stewardess answered confidently.

"But what if they check?" Dana protested.

"Do you want to get in or not, mate?"

Dana returned to Paul. "Just fill out 'the Los Angeles Days Inn,'" she whispered.

The ruse worked. Dana and Paul filled out the phony address, sneaked to the back of the line, and made their way forward to INS as if nothing had happened. The Customs inspector looked at the I-94 forms, checked that all the fields were filled in, and stamped their passport. Within minutes they were getting their luggage and walking to the cab line, where Phil had prearranged for them to be picked up.

Holding Pen

"About the apartment . . . " said Hector—a guy who worked Stateside for Phil and Ray—as they barreled down Century Boulevard, away from the airport.

"We know, we know," said Paul. "We're doubling up."

"Well, I hate to be the bearer of bad news," said Hector, "but my instructions are just to take you up to Irvine."

"What's up in Irvine?"

"That's where we take people who are waiting for apartments to become available."

"You mean, like a 'safe house'?" offered Dana.

"I don't think I'd call it that," said Hector, switching lanes without signaling.

"You don't happen to know about the car we were promised," said Paul.

"No, I don't."

An hour later, Hector had left them standing outside a beige-colored, box-shaped house on a quiet, tree-lined side street.

"This doesn't look so bad," said Paul, looking up toward the top of the palm trees that swished against a second-story balcony. Then his gaze fixed on a man, seated on the balcony, who was smoking a cigarette.

"You must be new arrivals," he said with a thick Russian accent.

Aliens: Twice the Work for Half the Pay

"Yes, we are," said Paul.

"Well, come upstairs then. Apartment 2G."

They clunked their luggage up an inside staircase and walked down a short corridor that led to the apartment. As they approached it, a dull, rumbling, thumping vibration seemed to be drawing nearer.

"Stop," said Dana, squeaking her rolling luggage cart to a halt. "Sshh. Is that an earthquake?"

"I don't think so," said Paul. "But it does sound like some kind of explosion—or some kind of gun battle. Maybe a sound effects record?"

They moved ahead down the hall, found the door of Apartment 2G, rang the bell, but nobody answered. The sound effects—explosions, gunfire, machine-gun rat-a-tat-tat, were now almost unbearably loud.

"Maybe you'd better bang on the door," said Dana.

"Hello!" said Paul, banging on the door three times. "Is anybody in there?"

The door opened. "Welcome," said the man with the cigarette. "Welcome to the Holding Pen."

"What's that racket?" shouted Dana as they moved through the foyer into the living room.

"The deathwatch du jour," the man said, ducking into a bedroom. "Singh, Steve, turn it down or put on earphones, please." He returned and shrugged. "It's like World War Three in there. Please, sit down and make yourselves pleasant."

"Thanks," said Dana, flopping back on a battered couch that was the only piece of furniture in the room.

Flophouse

"So you want the grand tour?" asked the cigarette-smoking man, who soon introduced himself as Vlad.

"Sure," said Paul. "I guess it would be good to know where we'll be sleeping."

"Well, probably right where you are. We have full house now. I am in small bedroom; Steve and Singh are in larger one. This is biggest remaining room."

"You mean we don't even get our own bedrooms?" asked Dana.

"This is holding pen," said Vlad. "Not Beverly Wilshire hotel."

"You can take the couch," said Paul to Dana. "I'll buy a sleeping bag."

"You'd better make it a waterproof one," said Dana, feeling the carpet. "This is wet."

"Yes, roof has problem with leak," said Vlad, pointing to a jagged crack in the ceiling. "Fortunately, not much rain in Sunny California. Come, I'll show you the bathroom and the kitchen."

"This is bloody encouraging," said Dana as they passed a laundry nook with a compact washer/dryer unit.

"Yes. Very useful," said Vlad. "But dryer doesn't vent well—another reason for dampness."

"Bathroom is in here," said Vlad, poking open a nearby door. "Sometimes works well, sometimes not. If jam occurs, by no means use Drano. Unclog manually, with plumber's helper."

"Do you have a phone?" asked Dana. "I really need to call Australia."

"The phone's in kitchen," said Vlad. He inched closer to Paul.

"What is wrong with woman?"

"I don't think she's ever been in a place like this before."

"Women are always pain," said Vlad.

Dana dialed the recruiter's Australian office, but nobody was around. She almost left an incendiary message on the tape, but thought better of it.

"Idiots," she said, replacing the phone in its cradle. She returned to the kitchen, where Paul and Vlad were sitting on two battered stools at a tiled counter.

"Do you guys mind if I clean out the refrigerator?" asked Dana, after opening its door and wincing in horror.

"Well, that's sort of a hot-button issue," said Steve, who reached around her to retrieve a beer from the lower shelf. "The general rule is if a food item has got a label on it, it's the property of the person who labeled it. Unless, of course, it's clearly past the point of no return."

"How far would you say this one's gone?" asked Dana, holding up a stained carton of chow mein whose top was ringed with mold.

"An obvious non-performer," said Steve.

"And this one?" asked Dana, waving an aluminum foil pan half-filled with an unrecognizable substance.

"A science experiment," said Vlad.

"Yeah, like something the KGB would have cooked up," said Steve, retreating to his game of Deathmatch.

"I think you guys think this is fun," said Dana, kneeling to retrieve a set of rubber gloves and a bottle of Formula 409 from beneath the sink. "Living like pigs, I mean. The experience must take you back to your college days."

"I didn't go to college, I was in army," said Vlad.

"But you're an H-1B aren't you?" probed Paul. "Isn't that for graduate and post-graduate workers only?"

"Recruiter in former Soviet Union fixed it for me."

Doing the Numbers

Over the next two days, Dana and Paul did their best to make their tiny space in the "holding pen" more comfortable. After learning that only one car was allotted to residents of the "holding pen," they borrowed it and drove to Costco, where Dana bought some bedsheets and Paul picked up a weatherproof sleeping bag. Dana tacked the sheet to the ceiling to divide the room, and Paul laid the sleeping bag out on the floor. Paul also bought a small white noise generator to mask the sounds which bedeviled the place at night: the sound of the dryer, which clunked rhythmically throughout the night, the clattering of late-night kitchen maneuvers, and the virtual gunfire of Quake.

Dealing with the apartment's occupants—Vlad, Steve, and Singh—was a matter of making as little noise as possible. The programmers worked in shifts: Vlad, from 7:00 P.M. to 7:00 A.M.; Steve, from 10:00 A.M. to 7:00 P.M.; and Singh, from midnight to 8:00 A.M. Their strategy was to spend as much time away from the apartment as possible, which was difficult, given the nearest mini-mart or fast-food restaurant was four miles away and the car, which was continually in use ferrying programmers to jobs or returning them after their shifts ended, was often not available.

The nearest all-night haven was Taco Bell; on Sunday night, Vlad gave them a lift there.

"I'll be glad to get to work tomorrow," said Dana. "No offense, Vladimir, but that apartment is the pits."

"No offense taken," said Vlad, chewing on a burrito. "So you've been assigned to InfoDelta?"

"Yes," said Paul. "Know anything about it?"

"I've worked there," said Vlad. "Not the worst company—not the best. Most people working there come from body shops. Like big factory."

"Body shops?"

"Like the one we work for. Each body—you, me, her—makes fortune for Phil and Ray. Tell me, how much does your contract say to pay you?"

"$5,000 a month," said Paul.

Aliens: Twice the Work for Half the Pay

Vlad shook his head. "I am always amused at the numbers."

"What do you mean?"

"We get billed out to InfoDelta—or wherever—at maybe $100 an hour. For sixty-hour week, that is $6,000. We get $1,250—Phil and Ray get $3,750. They pay for crummy apartment and crummy rental car—keep the rest. So, for four people, they make $15,000 per week."

"Pretty good money," said Paul.

"We're in wrong business," said Vlad. "Forget programming—think importing of people, bodies."

"Why can't you work for somebody else?" asked Dana.

"You kidding? Am here on H-1B. Switch job, lose status. Have ten days to leave country. Even with permission to switch job, green card application gets reset to zero. Like mouse sent to front of maze."

"What are you going to do?"

"Do my job. Keep quiet. Save money, maybe find my own place and car. Wait for green card. Then can stay and write own ticket."

"How long will that take?"

"For me, two more years of this."

"Yikes."

"I am single guy. Living in pigsty not a problem. Besides, better here than in former Soviet Union."

At around 11:00 P.M., the threesome returned to the apartment. A message from Phil was on the answering machine: "Hey guys—Dana and Paul—there's been a change in plans. No problem, really, just a change. Forget InfoDelta—at least for the moment. Report tomorrow to WorldData—it's in the next complex over. They'll have plenty of work for you. I'm going to be coming over next week to help sort things out. Maybe I'll see you both then."

"Christ," said Dana. "It's not even the same company."

"Vlad," asked Paul. "What do you know about WorldData?"

"Another big factory," said Vlad. "Not the best company, but not the worst."

We Are the World(Data)

On Monday—their first day of work—Dana and Paul got up early. After showering, some instant coffee, and some cold cereal, they were ready for work.

"Who's not working today?" queried Vlad at 7:30 A.M. "Steve—you're on the bench this week—want to drive them?"

"Okay," said a sleepy voice in the large bedroom. A half hour later they were on the road.

"Thanks for driving us," Dana said to Steve.

"No big deal," said Steve. "I didn't really have any plans today."

"Aren't you working this week?" asked Paul.

"No. I'm 'on the bench.' It happens a lot these days—but I can't complain. At least I get paid, which is more than I can say for other body shops out there. Take my cousin, up in Silicon Valley. He spent $5,000 for a bogus technology 'crash course' in India, got sponsored, and now he's virtually a slave. When he works, he gets paid maybe $1,000 a week. When he doesn't, he gets just $125 a week. Lives in a pen much worse than this one."

"Don't you guys ever get pissed enough to do something about your situation?" asked Dana.

"The way I look at it," said Steve, "the best revenge is to get rich. And I'm not going to get rich in India. Or at least, not as rich as I might get here."

"What are the odds of that happening?" asked Dana.

"Better than you might think," said Steve. "More than three quarters of the start-ups in the Valley are started by Indians."

A half hour later, Dana and Paul were dropped off at WorldData's looming glass-walled headquarters. Dana was first to meet with the hiring manager.

"Well, we're really glad you're here," he said. "We've got a big Java implementation project that we're crashing the time on for the next six weeks."

"Java?" asked Dana.

"Yeah," the hiring manager said. "Is there a problem with that?"

"No, no," said Dana, "but I thought I'd be doing database work."

"We've got all the Sybase people we need right now," the hiring manager said.

"Got it," said Dana.

The hiring manager briefed her on the details of the assignment, assigned her to a cube on the third floor, and sent her on her way. As she was passing Paul on the way into the office, she rolled her eyes.

"We're fucked," Dana said, showing him the briefing sheet.

"You sort of know how to write Java, don't you?"

"Not really," said Dana. "But if they don't care what I do, why should I care?"

She marched down the stairs to the third floor of the complex and pushed open a steel door. Inside was a massive room that was surrounded by green, anti-glare windows and was filled with about two hundred programmers. After getting lost twice in the maze of cubes, she finally found her space—cube 3786. Inside it were a PC, a desk, and an Aeron chair. It was spartan, but Dana was happy; the surfaces were clean and nothing smelled of rotting food or moldy laundry.

Charade

Dana spent the next week looking as busy as she possibly could, without doing any actual work. Doing this was easy. While there were meetings being held in an inner core of conference rooms, nobody seemed to know that she was there, nor did anyone come by to supervise what she *wasn't* doing.

"This is great," Dana said, when she caught up with Paul in the cafeteria.

"What have you been doing?"

"Oh, just answering the phone whenever it rings. Whoever was in my cube before me seems to have run up a bunch of debts. I keep talking to American Express."

"Hmm," said Paul.

"And then I call Phil and Ray, and ask them to get us out of that goddamned hellhole."

"I'm almost starting to get used to it," said Paul.

"Well, I'm not getting used to it," said Dana. "I mean, I know that because we're foreigners we're supposed to just shut up and smile and make the best of it, but it really doesn't suit me. The only good thing about this situation is the fact that I'm doing absolutely no work, but that doesn't make me feel good either. It just makes me feel like I deserve this kind of abuse—like I'm a cheater or a malingerer. You know what I mean?"

"Yeah," said Paul.

"So what have you been doing?"

"Well, I've been assigned to a Y2K project. But there's no actual bug-fixing involved. All we do is remove the stupid bug fixes that were put in a few months ago. Then, once those are removed, we hand it over to another team who will apparently install software that will remove all the bugs that we put back."

"How long will that take?"

"I'm not sure."

"Well I've made up my mind about one thing. If Phil and Ray don't do something about our living conditions in the next week, I'm going to quit. I don't care if I have to take a tramp steamer back to Sydney. I'm sick of living in that rat hole."

"I hear you," said Paul. "My clothes are beginning to really stink from it."

The Rainmaker

Another week went by, but despite Dana's repeated protests, Phil and Ray did nothing to extricate her, or Paul, from the holding pen. The refrigerator that Vlad had promised to clear out continued to reek and even the bathtub, whose drain was clogged with five different shades of hair, refused to drain, forcing Dana to stand in murky water while she took her morning shower. But the last straw was the rain, which came unexpectedly, in a violent deluge, drenching the carpet and part of the couch. In a last-ditch effort to limit the damage, Paul put out a bucket to catch the rain, but the noise of the water droplets was torturous.

"I'm giving these bastards an ultimatum," whispered Dana to Paul that night, as the programmers, safe and dry in their bedrooms, snored in the darkness. She did, leaving the following message on Phil and Ray's answering machine: "Phil, it's Dana. Sorry, but this living situation is wearing us out. We've been here three weeks, but we still don't have a car, or even a room, much less a livable space. Fix the problem—fix it now, or we're on the next plane back to Australia."

The next day—Friday—was business as usual. Dana was in her cube, pretending to do work; Paul was in another, removing work that had been done by another team. But when they returned to the apartment that night, they ran straight into Phil, who'd flown in the night before. He was clearly braced for a fight, but Dana preempted him.

"Well, Phil!" she shouted icily. "Can I get you a cup of mold from the fridge? Better yet—would you like a nice cold shower? Just stand over here!"

"What's wrong with her?" asked Phil of Paul. But Paul just shrugged his shoulders.

"Listen, Dana," Phil said. "I've got you and Paul booked into the next apartment that comes available. But it's not going to happen tonight and it's not going to happen tomorrow. You've got to swing with us on this, please."

"What about this stupid assignment you've got me on? It's not even with the same company you told me about, much less having anything to do with databases."

"It's temporary," said Phil. "Look . . ."

"You look," said Dana. "We didn't sign up for this kind of treatment. I told you on the phone—fix this shit now—get us what you promised us, or we're out of here."

Now the voices were loud enough to attract the attention of Steve and Singh, whose heads popped out of the large bedroom door. Even Vlad—who normally slept at this hour—had been roused by the commotion, and stood, yawning, near the foyer. Sensing that he now had an audience, Phil went on the offensive.

"Just who the hell do you think you are?"

"I'm a fucking human being. What are you?"

"Leave now and I'll sue your asses in court. You signed up for three months."

"Oh yeah? What about our work permits? We've been working here illegally for the last two weeks."

"I told you, we're taking care of it."

"Taking care of it—WHEN?" She turned to Vlad. "Tell me, Vladimir—what did he promise you to bring you over here? A house? A car? A green card?"

"This is not my problem," Vlad said, waving his hand and returning to his room.

"Steve, Singh—how about you?" she probed, but the two programmers simply ducked their heads back into the room.

"You're just a stupid, selfish cunt," said Phil. "If you're interested in being paid for your work, contact my lawyer. If not, you both can go to hell." And with that, he turned on his heels and slammed the door.

A half-second later, the bucket of rainwater that Dana had tossed crashed noisily against it.

"Well, I guess that tears it," said Paul.

The Last Breakfast

"Why didn't you guys stand up for me?" asked Dana, the next day, as they all sat around the kitchen table. "Don't you guys have any . . . any . . ."

"Honor?" offered Vlad.

"Yeah—fucking honor."

"Let me explain it this way," said Vlad. "You can make big noise, throw bucket, and go back to Australia. Australia not so bad. Maybe I could do something similar, lose job, and go back to former Soviet Union—it's bad, but not so bad as for Steve and Singh. You want to see hellhole? Try town where they're from."

"So this is your strategy—to simply trust these guys, lie low, keep your mouth shut, hope that things get better," said Dana.

"I don't hope," said Vlad. "I work, and I save, and I stay out of trouble, and I keep my mouth shut. But one day—you'll see. I will be in America—and you will still be wherever you come from."

"It's funny," said Paul. "Back on the plane, when we had to fill out those forms stating our purposes in coming here."

"The I-94s," said Dana. "And we put in 'Holiday Purposes.'"

"Yeah," said Paul. "And now here we are, in a strange house, in a strange land, without a paycheck, and we've just been told we're going to get sued."

"So you call this a vacation?"

"Only in the National Lampoon sense."

"I really like wife in that movie," said Vlad.

"I just want to know one thing," said Dana. "Am I a stupid, selfish cunt?"

"You want the truth?" said Vlad.

"Of course," said Dana.

"Well," said Vlad. "I'll say this—you're not stupid."

Epilogue

Dana and Paul returned to Sydney a few days after Phil had told them to go to hell. Doing so almost wiped them out financially; Dana had to borrow $1,000 from Paul to come up with enough money to pay her $1,400 ticket. Each returned to the Australian job market, where they now make decent but non-stellar salaries in the IT industry.

Vlad, Steve, and Singh are still guest workers in the United States and, despite the downturn in the tech economy, still employed. Like itinerant, migrant workers of yore, their powers of endurance, willingness to work cheaply, and immense talent for keeping their mouths shut continue to make them desirable employees, despite the anti-immigrant fervor stimulated by the events of September 11, 2001. Dana doesn't know whether they're still in the fetid "holding pen," but she doesn't doubt that, somewhere, they all have a roof over their heads, even if it's a leaky one.

— Pawns: Twenty Years of Typing and They Put You on the Web Shift —

Pawns: Who Are They?

Are you a Pawn? We hope not, because, if you are, you're going to have to get us both a nice cup of coffee, make sure the clients in the conference room have fresh bottled water, change the toner in the laser cartridge, and clear out our junk mail.

Hey, while you're at it, why don't you order yourself some flowers—we're getting close to National Administrative Professionals Appreciation Day—better yet, order them for all the secretaries. If you don't overspend, and you make triplicate copies of the receipts and send a nice corporate-wide e-mail so that the other executives realize how enlightened we are, we'll refrain from firing you today. Wait a minute—we can't fire you—without you, we couldn't even find our start button. So stick around, will you? After all, Kings like us need Pawns like you.

Pawns—in case you haven't figured it out yet—are the twenty-first-century den mothers (and fathers) who sought to bring some semblance of order to the oversized playpens of the New Economy.

For Pawns working at Net start-ups, life was particularly stressful. While everyone else was playing foosball and paddling around on Razor Scooters, Pawns were expected to behave like fully grown, rational adults. There was no late-night clubbing, long lunches, or elaborate, expense-account-fueled orgies at Nobu for them. They were expected to show up on time, sit placidly at their desks, process invoices, screen angry phone calls from creditors, and otherwise present a "professional" demeanor that belied the insanity that was happening around them.

Many Pawns shunned the start-up experience completely, preferring to stay at "stable" old economy companies where, if they were "good soldiers," and played by the rules, they would be taken care of and rewarded for their efforts with an undemanding schedule and a predictable gold watch after twenty-five years of loyal service. Unfortunately, however, there was no escaping the Internet's influence on the workplace.

Suddenly, it was no longer enough to be able to type 80 WPM, take dictation, run the boss's social and professional calendar, and keep a smile on your face from 9:00 to 5:00. Pawns were now expected to become virtual Swiss Army knives capable of managing Access databases, updating the corporate Web site, and performing emergency repairs on the boss's constantly crashing laptop.

The Internet also abolished the most important term of Pawns' traditional contract with their employers, the one specifying that all the humiliation, idiocy, and frustration of being a minion would end promptly and faithfully each afternoon at 5:00. In the New New workplace, Pawns had to cater to their bosses' whims 24/7—whether they were in Denver, Dayton, or Damascus.

For Pawns, and for the rest of us, the most important change wrought by the Net was in how it eroded the traditional paternalistic parent-child covenant that in olden times guaranteed that loyal, hardworking Pawns would always be taken care of.

Today, in the post-bust world of overnight mergers and constant corporate downsizing, Pawns—even those who've embraced the Net—don't have a lot to look forward to. When budgets are cut, and their bosses are sent packing, they're often the first to be cut loose. The ones lucky enough to survive inherit the workloads of their fallen comrades, and are expected to perform these additional duties without the least hint of dissatisfaction or ingratitude.

Are You a Pawn?

You might be a Pawn if . . .

- When you're not typing, filing, or changing a printer cartridge, you're forwarding dumb jokes to your friends or swapping emoticons via Instant Messenger.

- You know your boss better than his wife, his shrink, and his mistress (and if you play your cards right, you might wind up being one or all of the above).

- There's at least one national holiday that's been created by the floral industry to celebrate your sorry, shit-assed job.

- Your name is so far down on the org chart that you need to use special "skyscraper" paper to print it out.

- You're the only one in your office who looks forward to a software upgrade, not because it will make you more productive, but because it will allow you to finally kick that smart-assed IS guy in the ass while he's fumbling around under your desk.

- Unlike other NetSlaves, you never asked to be part of the Internet revolution. You were drafted one afternoon, while out Christmas shopping for your boss.

- You know more about the inner workings of the company than the CFO, the CTO, the COO, and the CMO combined, but the only thing that people ever ask you about is whether their take-out food has arrived.

- When your last boss got fired for gross incompetence, he was rewarded with a $20 million severance package. All you got—after twenty years of loyal service—was a miserable two weeks, and all the Post-it notes and Wite-Out that you could cram into your bag.

Fun Facts About Pawns

How Other People View Them: Gum-snapping, Garfield-reading, bubble-headed paper-pushers.

How They View Themselves: The only person in the goddamned office who actually does any work.

Favorite Movies: Nine to Five and *Working Girl*, along with *Pretty Woman*, *Erin Brockovich*, and any other vacuous movie starring Julia Roberts.

Former Pet Phrase: "I'm sorry, but Mr. Jones is away from his desk right now."

Current Motto: "Twice the work for half the pay."

What They Did Before the Internet Boom: Filing, faxing, photocopying, and walking the boss's dog.

What They Did During the Boom: FTP-ing, formatting, de-fragging, and walking the boss's dog.

What They're Doing Now: Being pulled in fifteen directions by five different bosses in three different time zones.

Favorite Songs: "Take This Job and Shove It," "Everybody's Working for the Weekend," "Bang on the Drum All Day."

Psychological Profile: Passive-aggressive. Most low-level admins realize that they can't directly confront the evil in their office, and therefore resort to underhanded and secretive means of exacting revenge. Favorite tactics include throwing out all of the boss's holiday cards instead of mailing them, pouring Visine in the coffee, and gossiping about the boss's latest sexcapades—real or imagined.

Pawns: The Story of Kaye

Kaye never wanted to be a secretary. She'd dabbled with many vocational pursuits: sound mixing, photography, a couple of video jobs, and even a brief stint as a punk rock music manager. But, time after time, these flaky self-employment schemes fell through, and Kaye was left to pick up the pieces, which meant getting another secretarial job. Then she had a kid, which meant giving up the one secretarial job that had some real career mobility: a demanding sixty-hour-a-week job at a prestigious Manhattan legal firm.

Thirty credits short of completing her college degree, she'd never found the time or the money to finish it. Now, at age thirty-eight, Kaye found herself at Reach Communications, a third-rate classified ad sales company that paid her a fourth-rate $28,500 a year to serve as an executive secretary. Hal, her boss, was a guy in his fifties whose concept of what an executive secretary did included personal errands, dogwalking, and shopping for his wife. Jake, his partner, was downright brutish. He'd complain about the coffee, make off-color jokes about Kaye's dress, and behave like a general asshole most of the time.

"You can do better than this," said Em, Kaye's younger sister, who'd recently joined the IT world as a graphics designer.

"I can't do what you've done," Kaye confessed. "I can't just take time off to reinvent myself."

"Your company won't retrain you to learn the Web?"

"This company is paper-based. The partners don't care if the company goes down the drain in five years. They'll be long gone."

"You should get yourself out of there. Look, I can get you a job where I'm at, but you'll be working around the clock—it's a start-up. And anyway, you're really not a geek."

"I'm glad I'm not a geek," Kaye said. "You people work around the clock. You also never seem to bathe."

"Yeah, well, I became a geek and I tripled my salary in three years."

"You're a designer—a bohemian. I'm just a middle-class matron. A typing slave."

"Get some training," said Em.

"I don't have the support system. I have to raise a kid," said Kaye.

"Do it at night," said Em.

Last Ass to Wipe

Kaye heeded her sister's advice. She paid her mother $50 to watch the kid for two nights a week, and started taking courses at the Learning Annex. She tried to get her company to pay for a course in FrontPage, but they gave her the cold shoulder.

"I'm not going to pay for you to learn something I don't know," said Hal, throwing the expense application back on her desk.

"Fine—I'll do it on my own," said Kaye.

She did. After breezing through thirty hours of basic FrontPage, PageMill, and HTML classes, she began immersing herself in the arcana of PERL, FTP, and other Net tools. It was tough, but more intellectually challenging than anything she'd ever experienced in her 9:00 to 5:00 work life. The Web, when you got right down to it, was a lot less complicated than memorizing the control keys of WordPerfect, WordStar, Vydec, Wang, or the Mergenthaler Linotype Omnitech 2000 Document Production System—outdated technologies that Kaye had mastered long ago.

To test her newfound skills, Kaye began to fool around with her own Web page on Geocities. She soon upgraded the site and began publishing her essays, which ranged from treatises on Stockhausen to minireviews of unknown artists distributing their wares on MP3.com.

"I didn't know you knew how to write PERL," marveled her sister.

"I don't. It's Flash."

"You know, I can probably get you a job at a start-up next week."

"Is it the kind of job where I can come in at 9:00 and leave at 5:00?"

"No—it's a start-up. Come in at ten. Work until whatever."

"Well, I'm willing to be exploited and shat upon by morons, but only for forty hours a week."

"Maybe you should take some anger management courses before you do anything," said Em.

Résumé in Play

By mid-2000, having endured three humiliating years at Reach Communications, Kaye finally began marketing her updated, upgraded résumé on Monster.com and HotJobs.com. In those gung ho, pre-NASDAQ crash days, Kaye found many takers among New York's insurgent New Media industry. The Net might have changed all the rules, but CEOs—even the brashest, youngest, peach-fuzz-faced ones—needed obedient functionaries to make their coffee, keep their appointment books, walk their dogs, and wipe their asses. Kaye could have walked into a $65,000 job overnight, but was appalled by the spartan conditions of these downtown companies, whose crowded offices were housed in tiny lofts and former garment district sweatshops.

"What is it with these people?" Kaye asked her sister, who, by 2000, had worked for at least seven different New York start-ups. "The last office I was in looked like it had been hit by a bomb. The CEO was actually sitting on a milk crate when he interviewed me."

"Lean and mean is in," said Em. "Frills are out. It's all part of the new dot.com paradigm—do more with less. Get rich by looking poor."

"Sheesh—I thought these were real companies."

"On the Internet, nobody knows that there's no heat in your office."

"Well, I hope there are a few companies out there that pay their utility bills."

"Buck up," said Em. "Your résumé is in play now. You'll find something."

In September, she happened upon Time Warner's online career database—a public intranet run by the giant media company. Buried among the thousands of listings was an ad for a corporate secretary with a salary of $45,000 a year. Kaye, after tweaking her electronic résumé, submitted it, and about two weeks later—without the intervention of a personnel agency, recruiter, or personal career trainer—got an appointment with Time Warner's HR department.

Just after Election Day, Kaye decked herself out in her best, most understated clothes and walked through under the massive Time Warner logo to a waiting elevator car that whisked her to the twenty-eighth floor. After clearing security, she was ushered into a conference room—the grandest one she had ever seen.

"Your résumé is impeccable," said Florence, a woman in her late forties who would be Kaye's immediate boss. "And your references check out. The fact that you know all about the Internet sort of clinches it—we're looking for people here who really know about the power of online."

"Well, the whole world is converging," said Kaye "And you're in the middle of it."

"We sure are. And if you come to work here, you'll be in the middle of it too. Mr. Danton—that's who I work for—is one of Mr. Levin's most trusted lieutenants."

"So would I be working for you or Mr. Danton?"

"Well, for me *and* Mr. Danton," Florence said. "Senior executives on twenty-eight rate two secretaries—a primary and a secondary. Your position is for a second, somebody who helps me help Mr. Danton. I cover 9:00 to 5:00, the second covers 1:00 to 8:00."

"Wow—that's some coverage."

"The Masters of the Universe like efficiency," said Florence with a wink.

"I bet they do," said Kaye, winking back.

In late November, after a round of reference-checking and a meeting with Mr. Danton—a crisp, well-groomed, good-looking man in his late thirties—Kaye received notification that she'd gotten the job. She was installed on the twenty-eighth floor outside Mr. Danton's office—a corner space with a gorgeous view of the Rockefeller Center Skating Rink. Outside this office—the plushest one that Kaye had ever seen—were two common work areas. Florence was at the front (where most of Mr. Danton's visitors entered and exited), and Kaye sat at the rear, near a door that was nearer to Danton's desk. To her left was the entrance to the same grand conference room that she'd had her interview in; to her right, a staircase that led up to the offices where Levin, Parsons, and other Masters of the Universe had their offices.

"You should *see* this place," said Kaye to her sister when they next talked. "It's like I've died and gone to heaven. The elevators are gilded, Em—*gilded*. My boss is a dreamboat and I've got a bunch of free tickets to see *Castaway* at the Ziegfeld—would you like to come?"

"It sounds to me like you're awash in perks," said Em. "But I can only wonder what Paul would make of it."

"Paul?" said Kaye, invoking the name of her ex-husband, who'd worked for Time Warner in the late-1990s. "Please, don't remind me of that loser."

"Okay," said Em. "So tell me more about your dream job."

"Sshh," said Kaye "Mr. Parsons is coming down the staircase. I've got to run."

Soon, Kaye's daughter's sunny little room was bedecked with a small army of fuzzy, branded WB characters. Kaye spent her lunch hours at MoMA, at the Warner Bros. Store, or at 75 Rock's plush cafeteria.

It was hardly a dream job—her duties, when you thought about it, were pretty mindless. She kept Mr. Danton's appointment book, his calendar, and parsed his e-mail box, which piled up with all sorts of group goo every day. She made sure she knew exactly where he was for every moment of the workday, just in case Levin or Parsons or some other honcho demanded his immediate attention. She ran his life in the same way that a dogwalker guided a Doberman, using Outlook, a RIM Blackberry, and strategic text attachments in lieu of a leather leash. But working at Time Warner in its heyday—when the kitchen commissary was always stocked with sandwiches, and there were always free DVDs, CDs, and free Time Inc. magazines to take home—was the best secretarial job Kaye had ever had, bar none.

Stairway to Levin

Over the next two months, from her perch by the foot of the stairs, Kaye watched Levin, Parsons, Bressler, and other Masters glide from oak-mahogany meeting to oak-mahogany meeting. Any initial sense of being overwhelmed by being in the presence of the Media Gods quickly passed; awe passed to deference, and Kaye's workload was light. The office atmosphere was relaxed, almost philosophical in flavor—a clear debt imprint of Levin's quasi-collegial style of rule. The Philosopher King rarely descended from his upstairs lair. More often than not, Kaye's boss walked up the stairs, and back down, when an audience was concluded. When Levin did descend, to preside over a board meeting in the conference room, everyone on twenty-eight stood at attention.

Kaye had never worked as a "second" before. Her past bosses—even the highest profile attorneys or corporate presidents—never rated more than one secretary apiece. Most—like

Hal and Jake—shared a secretary; other unlucky secretaries worked for as many as four bosses at once. At Time Warner, however, Levin's central staff was so important that each rated at least two (with Levin having command of a personal staff of six).

"It's funny," said Kaye to her sister. "Everybody comes in here on time, and sits purposefully at their desks. But I've yet to see anybody actually do any work around here."

"Time Warner is sort of like a firehouse," said Em. Everybody sits around reading magazines for weeks, until the alarm bell rings. Then everybody bounces off the walls like Super Balls. Trust me—it'll happen."

"Well, I'm ready for when it does," said Kaye. "I'm sick of reading about Pokémon and Tom Cruise."

On the morning of January 10, 2000, the alarm bells finally rang on the twenty-eighth floor of 75 Rockefeller Plaza. "We are creating the operating system for life in the Internet Century," said Levin, collarless, to the crowd. "The opportunities are limitless for everyone connected to AOL Time Warner," said Levin to scattered applause, and everyone went back to their desks.

Within minutes, calculators were ticking throughout the office. Everybody was figuring their worth—their piece of the big $300 billion piñata.

The Rogues from Reston

"I still don't know what it all means," said Kaye, "besides the fact that the new company will be 'the operating system for life in the Internet Century.'"

"How is the twenty-eighth floor taking it?"

"There's plenty of chortling. And there's paper tape from the calculators all over the floor."

"Have you figured out your net worth yet?"

"I've only been here four weeks," said Kaye. "I'm not vested or anything."

"Too bad."

"I just don't want to be restructured right now," said Kaye. "This is the first job that I've actually liked doing."

"But that's because they hardly make you do anything."

"I do a lot," said Kaye. "I know where my boss is twenty-four hours a day—every minute of the day. Jerry Levin could call at any time. . . ."

"Well, at least you don't have to walk his dog."

"Amen, sister."

Termites in the Mahogony

It wasn't long before changes began to be felt at 75 Rockefeller Plaza. Within a day after the announcement, the dress code changed radically. Neckties were banished. Copies of *People* magazine and *EW* gathered dust on desks, as everybody—from $300,000 Master of the Universe to $60,000 secretary-factotum—pored over the latest analysis of the merger from *Barron's*, the *Wall Street Journal*, and the *New York Observer*. Executives, playing "musical chairs," were blabbing more to reporters than they were to their own subordinates. Things weren't sleepy anymore—everything was in motion—old ways of doing things were falling, and the new guys were in town.

The AOL guys began arriving in late January. Kaye quickly learned to recognize an AOL guy from a hundred paces. *They* were the suits—the efficiency experts—the men from the Internet lagoon. Their numbers expanded more quickly than termites in a Florida frame house. First confined to a single drab office on the thirtieth floor, the AOL guys soon moved throughout 75 Rockefeller Plaza. They demanded offices whose furnishings were equal to or better than those of TW's executives. Unfortunately, there wasn't enough space to accommodate more than a few of them (until such time that Time Warner's staff was massively reduced—an event that few doubted would occur as soon as possible).

"It's the end of Steve Ross," wept one career secretary who'd been at TW since 1978.

"I just wish they'd use drop cloths," complained another, dusting her shoes of the dust from the Sheetrock sanders.

Dancing on Eggshells

In the meantime, the media was crawling all over the long-term ramifications of the Time Warner merger. Some—like Tom Brokaw—considered it the greatest thing since the invention of fiber optic cable. Others likened it to the Hitler-Stalin non-aggression pact of 1939. Down in Washington, AOL lobbyists greased the merger's chance of proceeding without a hitch with whispered assurances, firm handshakes, and soft money. But others in government—including Jesse Helms and Orrin Hatch—noisily pledged to investigate the merger's long-term ramifications. At the same time, the FTC was pondering its anti-trust implications, and the FCC was gearing up for a round of hearings for the summer.

Everyone was walking on eggshells. The world was hanging in the balance. But then a huge fuckup occurred that threatened to derail the whole deal.

One morning in early March, 13 million Time Warner cable subscribers woke up to a world that had been purged of Walt Disney's programming, including the entire ABC network (which carried such popular shows as *Who Wants to Be a Millionaire?*). With the simple flip of a switch, Joe Collins, Time Warner's notoriously pugnacious head of Cable Operations, had banished the Disney Empire—something that most Time Warner executives had secretly wished, hoped, and privately prayed for.

But TW's long-suffering subscribers had had enough. They raised hell—called the press, made angry phone calls to the Public Service Commission—the works. It was a shitstorm.

In the thirty-nine hours that Disney went dark on Time Warner's network, a complicated story of arcane contracts emerged. TW's legal position—that Disney was trying to strong-arm it into carrying additional programs—might have been impeccable in court, but in the court of public opinion, it was a disaster. Everybody—from the *New York Times* to the FCC—was talking about "monopoly power running amok." It was the worst possible fuckup at the worst possible time.

In any ordinary company, Joe Collins's head would have been served up on a plate to Steve Case, who was so infuriated by the gaffe that he called it "the dumbest move I've ever seen." But Collins had powerful friends, a fact that made him "untouchable"—within the organization.

So the order came down: Find a sacrificial lamb, a fall guy, and that person was Kaye's boss, who suddenly lost out to his AOL counterpart, an executive who'd recently been relocated to New York. The next day, both Florence and Kaye met with Craig Miller, who obviously had a different view of secretaries than Mr. Danton.

"I'll be frank with you," Miller said. "I've never used a secretary—I've always done my own typing, filing, faxing, etc. I keep my own calendar, make my own appointments, and frankly don't like sharing any information with anyone whom I don't completely trust. Florence—you can stay where you are for now. But I have no need for two secretaries—in fact, I don't know why anyone would."

"So you're not going to need me anymore?" asked Kaye.

"After Mr. Danton's affairs are put in order, I expect you to find a new position ASAP."

Mr. Career Survivor

That night, realizing that she was trapped, Kaye finally called her ex. Once a high-flying content producer at Pathfinder, Time Warner's doomed megaportal, Paul had been unceremoniously fired for alleged "improprieties" resulting from an elaborate prank involving Walter Isaacson and Jerry Levin. The net impact of this was to further convince Time Inc. capo Don Logan that Pathfinder was a dangerous playpen for Internet-based mayhem that should be closed down as quickly as possible. Paul's career had never recovered from his misstep, and he wasn't faring well lately. After burning through two Soho-based start-ups, he'd started drinking again, resuming a habit he'd managed to control during their tumultuous eight-year marriage.

Kaye felt sorry for Paul—he had gone from rich media hero to burned-out zero faster than anyone she'd ever known. Once tanned and athletic, he'd become gaunt, his hair had whitened, and his efforts to grow a beard that never became more than an extended stubble made him look like a homeless cart-pusher on Sixth Avenue. Paul's physical metamorphosis had been accompanied by a mental change. Once optimistic, enthusiastic, and jazzed about the ubiquitous interactive future, Paul had become a cynic, a critic, a Time Warner basher convinced that everything the organization did was the result of a mafia-like conspiracy.

Paul had just returned Kaye's daughter from a visit to the Bronx Zoo when she ushered him into her apartment and explained her situation.

"Sounds like you're about two weeks away from a lay-off," said Paul. "If you refuse the typing pool assignment, they have the right to get rid of you immediately. So you've got to make it impossible for them to fire you."

"Okay, Mr. Career Survivor—how do I do that?"

"Is there anything you know that they want to keep quiet?"

"Well, maybe the fact that Jerry's the worst-dressed CEO I've ever seen."

"No good. *Wired* ran that story last month. Think of something substantial. Something that could really crater the merger. I mean, the whole world is hanging in the balance. Disney, the Open Access people, even Ralph Nader and Bill Gates are having conniptions. They see this thing not as 'the operating system for the future,' but as the marriage of Gangsta Rap to an online service that's not exactly beloved by everybody, especially Microsoft."

"I'm not going to start leaking inside stuff to the press."

"You'd be the only one on twenty-eight who wasn't. Back when I was at Pathfinder it was the first thing that you'd do after you checked your e-mail. Everybody was leaking then—it was the only way to fight against the people upstairs who were undercutting us."

"Did they do anything?"

"Oh, yeah. They sent up a bunch of guys from Virginia, and they installed a sniffer program to ferret out leaks. It was all based on code words, you know, so that if anybody used the term 'Wabbit' or 'Linerunner' a little guy across the street would read the e-mail, and vet it."

"Yikes. Sounds like a police state."

"It was child's play. Once people figured out that their e-mail was being read, they'd simply make phone calls. And they'd use the e-mail channel to spread deliberate disinformation—you know, 'Logan's head is on the block,' or 'Sagan's history.' Most of it was fake."

"Did they tap your phones?"

"I don't really know."

"Tell me, Paul. Why did they fire you? You must have really done something to piss them off."

"I really had nothing to do with it."

"Yeah, but you know about it."

"Yes," said Paul. "You see, there was this guy, Ari, who worked at Pathfinder as a secretary—a really strange bird. Everybody thought he was gay but it was really just an act. This guy was, well, he was just strange in a way that no secretary I've ever seen was strange. He'd worked all over the Time Life building—he'd temped on thirty-four, the floor where Don Logan and the other top editors work—and he'd learned how to imitate everybody, even Jerry Levin."

"He did impressions of Jerry Levin? How's that possible—Levin isn't exactly the most distinctive-looking CEO."

"Voice impressions. On the phone, Ari sounded exactly like Levin. So one day—and we bet him $500 he wouldn't do it—he called up Walter Isaacson, the head of *Time*, and said, 'Walter, this is Jerry. We have a crucial matter that you're required to attend to. Be at my home in an hour—or sooner, if you can.' Of course, Levin lives on the tip of Long Island. So Walter Isaacson gets into a limo, is driven out there, and when he gets there, the place is empty."

"Oh my god."

"Well, it was bad—really bad. Walter missed a phone call from Henry Kissinger, and there was a big brouhaha about it on thirty-four, and then, of course, they called in the goons, and we all went under the lamps, and I confessed. Ari kept his job—amazingly—and we all got by with a warning. But it gave the whole division a black eye, and we were all let go a few months later. People on thirty-four didn't take Pathfinder seriously after that—they thought it was some kind of loony bin."

"You know, I always knew you were stupid, but tonight, Paul, you've truly amazed me."

"It did happen," said Paul. "But let's forget my tale of woe. Let's talk about you. The way I see it, you've got to find a powerful friend, or make them afraid of you. Now I'm sure there are some documents that cross your e-mail inbox that your bosses don't want appearing on the front page of the *Wall Street Journal*."

"I don't know anything. Really."

"I doesn't matter what you know," said Paul. "It only matters what they *think* you know."

Knife in the Back

Kaye dismissed her ex-husband, repelled by his insistence that her best chance of career survival lay in committing multiple acts of treasonous corporate espionage. But she also knew that the clock was ticking—she was due to receive an "offer that she could not refuse" from the HR people within days, and Florence had already taken steps to arrange a lunchtime meeting with the head of the legal department. Kaye realized that her boss meant well—a job, after all, was still a job—but took it as a knife in the back.

As the annual tree lighting ceremony took place in Rockefeller Center, Kaye began to openly doubt her chance of lasting through Christmas. Miller didn't want her, Florence was conspiring to get rid of her, and the legal department—which supervised the typing pool—was lusting to lure her into certain oblivion.

Kaye should have been dead meat. But fate, luck, and the peculiar circumstances surrounding the AOL–Time Warner merger conspired in her favor. One day, after auction-surfing and trading e-mails with other auction junkies, she found out about a Van Gogh painting that had come up for private sale in Holland.

Kaye came across many exotic items for sale on the Web. But when she learned about the Van Gogh, she immediately thought of Jerry Levin. Levin had no idea who Kaye was—but

Kaye knew he appreciated fine things. One of the main perks of being at Time Warner was being able to obtain free entry to the many cultural institutions rewarded by Levin, including MoMA, the Met, and the Cloisters.

So she typed up an email summarizing the painting and sent it to Jerry Levin's secretary, complete with a 640 x 480 GIF.

Ordinarily, such a message—whose header read simply "Would You Like to buy a Van Gogh?"—would have harmlessly passed through the e-mail gateway without further incident. But, because it contained the words "Van Gogh," a code word for a secret acquisition target—one with profound antitrust implications—it was immediately intercepted and routed automatically to the AOL inbox belonging to a technician in Virginia.

"You've Got Mail," said the automated voice.

"What the fuck?" answered the technician, quickly scanning Kaye's message and punching a phone. "Hey Charlie," he called to a comrade.

"Eh?"

"I've got a message here from someone in New York about Van Gogh."

"Van Gogh?"

"Uh huh."

"Forward it on to me. Wait—don't forward it on to me. I'll be in your office in five minutes."

"This can't be happening," said Charlie five minutes later. "Nobody in New York knows about Van Gogh."

"What the hell is Van Gogh?"

"I could tell you, but then I'd have to kill you," said Charlie. "Wait a minute—there's an attachment."

"Looks like a picture" said the technician.

"Open it up."

"I can't read it—it's corrupted."

"Damn. Why couldn't she have sent us an ART file?"

Pawns: Twenty Years of Typing and They Put You on the Web Shift

"What?"

"Nothing. Just forward it on to Steve," said Charlie. "If this is a leak, he's going to hit the ceiling. Oh, and delete the original message from your hard drive."

"Okay."

Within moments, Kaye's e-mail to Mr. Levin was being scrutinized again—this time, on an AOL-owned RIM Blackberry in an aircraft 30,000 feet over Delaware.

"Who the hell is sending out messages about Van Gogh?" asked a voice from the AOL Plane.

"She's just some secretary in New York," said Charlie. "Works for Miller—at least I think she does. I did a Web search for her—she publishes some kind of nutty e-zine."

"Christ," said the voice. "Get me Miller on the phone—can you connect him up?"

"Yes sir," said Charlie. Then Miller, in New York, came on the phone.

"Miller," said the voice on the AOL plane, "we've got to contain this thing—at least for a couple of weeks. I want you to keep an eye on this woman—she might be the source of the leaks to Disney. Last week, they got documents on Miro, Manet, and Pissarro. But if these fucks find out about Van Gogh and blab it to the FTC, well, I don't have to tell you about what the consequences could be."

"Yes, sir."

Van Go-Go

Miller slammed down the phone. The phone call made him furious. He, a supposed "insider," had never heard of the code word, "Van Gogh." Could this be something that Case and Levin were cooking up on their own? A reverse coup against Pittman, Novack, and Miller?

"Come into my office," Miller said to Kaye, gesturing inside.

"All right," said Miller. "What the hell is this 'Van Gogh' business about."

"You mean the painting that I sent an e-mail to Mr. Levin about?"

"Right," said Miller, his face twisting into a grimace. "The painting."

"Well, I'd really like to wait to hear back from Mr. Levin before I offer it to you."

"Did Mr. Levin get back to you about the painting yet?"

"No, I don't think his secretary's showed it to him."

"Did you send the e-mail to anybody else?"

"No."

"Okay," said Miller. "That is all—for now."

"It's $3 million," said Kaye, "if you're interested in it. But it would probably be better, politically, if Mr. Levin got first dibs."

"First dibs?"

Kaye left the office, confused, and resumed her game of Solitaire. Miller shut the door, and called Case's hotline in the air.

"It's a painting. Somebody in Germany told her about it."

"That's hardly believable," said Case.

"I'm not sure if she's just a social climber or if she really knows something."

"Did she mention any other artists?"

"No."

"Well, we can't take any chances," said Case.

"Don't worry about her," said Miller. She'll be off the executive suite in two days. She's going to be sent far away—down to the typing pool."

"Typing pool? That's where most rumors begin, you moron. Keep her there, somewhere where she can be watched."

"Okay," said Miller.

"Okay," said Case. "And for God's sake, Miller—keep this thing under your hat."

"Yes sir."

"And call me if she mentions any other painters."

"Okay."

So Easy to Lose

Kaye went home that night, a bit surprised about how quickly news of her offering of the Van Gogh had spread throughout the organization. Could it be that AOL people cared more about culture than suggested by their obsession with subscriber statistics and pop-up boxes?

That night, technicians requisitioned a complete copy of all the e-mails she had sent through the Time Warner e-mail gateway, and installed a bug on her office phone and PC.

Then the order that Florence had promised would soon arrive from the head of People Services—the one directing her to report for work in the typing pool—mysteriously got lost in the electronic ether. A week passed, and then another week, and another. If things went on this way, Kaye would survive her one-year period, and sail on to Golden Parachute land, where magic buyouts cushioned every fall from grace.

"It's funny," said Florence. "I've been here fifteen years and this is the first time that the People People have messed up a job reassignment this badly. They're usually quite prompt."

"Maybe it's the new AOL e-mail system," said Kaye. "It keeps crashing my computer."

"It's strange. But I'm sure you'll get your transfer soon— you must be going nuts, with nothing to do all the time."

"I get along. Surf the Web. Run my zine."

"Well, as it's okay with Miller it's okay with me," said Florence, with a wink.

"Mr. Miller is a pussycat," answered Kaye.

Soon, she had her sister on the phone again.

"It's funny, Em. I don't know what I did, but it looks like they're going to keep me here—at least until the time the merger goes through. Then I'm automatically vested, which will qualify me for an A-level Accelerated Retirement plan—the same that the big guys are getting."

"Wow—that's amazing."

"And my boss—the awful AOL guy I told you about—he's totally changed. It's funny—at first, I thought he was some kind of sociopath. He just stayed in his office, hunched over his notebook, and typed and typed. He wouldn't even let me make him coffee. But now, we seem to talk every day—he asks me all of these questions about art—you know, he's really a very nice guy inside. Maybe I was wrong about these AOL guys. Deep down, you know what—they're just regular people."

The Time Warner–AOL merger was approved on December 14, 2000. The very next day, Kaye was personally called into the president in charge of People Resources. He thanked Kaye for the "great job" she had done, offered her a six-month, $27,500 severance package, and made her sign a nondisclosure agreement that forbade her to discuss any matters relating to Time Warner for one year. Florence made her a wonderful going away gift—all the remaining glassware in the Time Warner Executive Pantry, and a Bugs Bunny pillow that used to adorn Mr. Danton's Eames Chair.

Epilogue

Kaye used her $27,500 severance payment to go to Germany to live for six months. There, she studied pursuits that had always concerned her—art, architecture, and music. When she returned to the United States in the fall of 2001, she was unable to find a job for three months, but then scored a decent position as an executive assistant to an accountant whose main practice is celebrity tax returns. Her strong suit—the factor that made it possible to stand out against many strong competitors, some with advanced degrees—was her glowing recommendations from Craig Miller. With a 9:00 to 5:00 job that pays her $80,000 a year and limited mandatory overtime, she's doing okay, and may complete her degree—perhaps at the University of Phoenix, a Web-based University.

She still doesn't know what saved her ass at Time Warner. Her sister believes it was due to bureaucratic inefficiency. Her husband blames it on flaky AOL software. But only those technicians deep in the heart of AOL's server cages know the answer. Each of them, like Kaye, has agreed never to comment on this incident, or any matter relating to the internal workings of AOL Time Warner, especially when it comes to matters of art.

— Bootstrappers: Forty Megs and a Stool —

Bootstrappers: Who Are They?

Bootstrappers are the happy misers of the Internet industry. As their name suggests, Bootstrappers run tiny but profitable technology companies on a shoestring budget. Being fiercely independent and distrustful of other people, Bootstrappers have never even considered accepting outside funding and view those who did during the dot.com boom as "a bunch of suckers."

While it's easy to criticize Bootstrappers for being small-time and somewhat paranoid, they certainly seemed like bearish visionaries when the market caved in. Such a view is, of course, exaggerated because Bootstrappers are bottom line-obsessed lone wolves whose approach to business would probably be the same no matter what the economic environment. In a sense, Bootstrappers are the crusty mom-and-pop storeowners at war with the local mall.

While the other categories of individuals we have profiled in this book fought against the Bubble's legacy of abuses through violence, litigation, or, indirectly, by opting out of the madness entirely, Bootstrappers have chosen "to rule in hell rather than serve in heaven." In our age of impersonal global corporations, it's comforting to think that there are people out there who don't want to sell out to the highest bidder and take great pride in serving their communities as small Internet Service Providers (ISPs), Web design firms, and Application Service Providers.

Even if you don't buy our romanticizing of Bootstrappers, the fact of the matter is that with all the stupid money long gone from the technology business, you'd better start acting like one or face the consequences. "Think different"? Hah! These days it's more like "Think cheap," "Think small," "Think really, really hard." And while you're at it, you can forget about becoming an overnight CEO sensation. You'll be lucky to be still in business after a year, never mind looking up Maria Bartiromo's dress while you spout off about your IPO.

Please excuse us if we sound like your grandfather who walked uphill, both ways, barefoot, in the snow, just to make it to that unheated schoolhouse. It's just that the situation couldn't be more different than what we have become accustomed to for the past five years.

For Bootstrappers, who at this point could be anyone who runs a company small or large, "the road ahead" (to use Gates's term) will be long and hard and ultimately belong to the scrapper who watches every penny and bends over backwards for customers. (Radical thoughts indeed for an industry notorious for putting the customer last, wasting billions of dollars of investors' money, and foisting junk products and junk IPOs on the unsuspecting masses.)

What to do? Just do the opposite of what everyone else did during the Boom, and then *maybe* you'll have a chance. Anything less, and you're toast, pal. Do you have what it takes? Turn the page.

Are You a Bootstrapper?

You might be a Bootstrapper if . . .

- People don't call you a "Bootstrapper;" they call you a "cheap bastard."

- Your idea of "lean and mean" would give most former free-spending dot.com managers nightmares. (For instance, "perks" at your company consist of heat, electricity, and a non-broken chair.)

- You bought all your office equipment at dot.com liquidation auctions.

- "Why pay more?" isn't your battle cry. It's your mission statement.

- You've outsourced development to shady Tech Sweatshops in Russia and India, where coders are paid $10 an hour. (You want the "new, new thing"? This is it!)

- You're running NT 4.0 on a token ring network. (For you non-geeks out there, this is the technological equivalent of a hamster running on a squeaky wheel to supply power to Detroit.)

- You pit wide-eyed, cash-starved vendors against one another to see who will slash prices (and their throats) the most to satisfy your pathological obsession with getting the "best possible deal." ("Vendor X said that they can do our corporate identity, and marketing, and build us a full-service e-commerce solution for $15,000. Can you top that?")

- "Mean Mr. Mustard" is your favorite Beatles song. ("Mean Mr. Mustard sleeps in the park, shaves in the dark, trying to save paper.")

- You do all your online shopping at e-wholesaler Overstock.com, not Amazon.

- The last time you bought a new computer was the first time.

- The closest you ever came to venture capital was borrowing $20 from an employee to buy sandpaper-grade toilet tissue for the company bathroom.

Fun Facts About Bootstrappers

Where They Can Be Found: Wherever there's a sale. Geographically speaking, however, our research on Bootstrappers suggests that they tend to congregate in small towns off the beaten track that are seldom known for technological expertise. (And are never "hip" enough to warrant such pretentious monikers as "Silicon Prairie" and "Digital Coast.")

Favorite Pastimes (During the Boom): Buying in bulk, shorting tech stocks, scoffing at the outrageous spending practices of the Internet industry.

Favorite Pastimes (After the Bust): Buying in bulk, shorting tech stocks, scoffing at the outrageous spending practices of the Internet industry.

Psychological Profile: 75 percent off.

Personal Habits: Domestic beer (in cans), dating people who don't mind "going Dutch," watching cable TV (thanks to their illegal descrambler box), renting movies from the library.

How They View Employees: The people to whom they pay the least amount of money for the greatest amount of labor.

What They Did During the Web Boom: Many Bootstrappers were penny-pinchers from birth. The remainder acquired their cheapness after getting laid off from companies that wasted billions of dollars on useless Super Bowl commercials and other nonsense.

What They Did After the Boom: Hoarded money, gloated.

Current Situation: Desperate. Even with their bottom-line, cost-cutting approach to business, many Bootstrappers are finding it very difficult to keep their tiny companies afloat.

Bootstrappers: The Story of Sally

Sally had always admired her older brother Jason. He was an extrovert, an athlete, a good-looking kid who never had a problem winning friends and influencing people. Sally, four years younger, was quieter, more inward-looking, and more likely to be found hiding behind a book than on a stage addressing an elementary school assembly.

Sally never came close to flunking any subject, but she never came home with any gold stars either. Her standardized test scores were good—in the 91st percentile—but not in the 99th.

In high school, she enjoyed a brief wave of popularity, but only because her friends knew that she could put in a good word for them with Jason (the captain of the football team). Somewhere along the line, she got the message: She was a good human being, but not a great one, and her future probably lay somewhere in a kitchen, not in a boardroom or an executive suite.

Still, Sally admired her brother, and always wanted to see him do well. She cheered louder than anybody when, at eighteen, he was sent off to Boston to attend Harvard on a scholarship. She listened patiently whenever her mother talked to friends about Jason's achievements: captain of the debating team, captain of the sculling team, and later, a writer for the Harvard Business School's newsletter. She went to college herself to study Art—a subject 180 degrees out of phase with Jason's pursuit of Economy and Global Finance. She painted but was too shy to show her work—to make ends meet she worked part time in an art gallery.

Jason was away in Boston for six years. He sent letters back to Northborough once or twice a month, but the only time that Sally and her mother actually spent time with him was on holidays.

Jason's Lament

It was on one of these Thanksgivings—in 1997—that Sally's world changed forever. Her mother, after clearing the last plate from the table, retired to the upstairs bedroom, leaving them to wash the dishes, wrap leftovers, and later, perform an old sibling ritual—smoking a homegrown joint on the back porch beneath a sparkling night sky.

"You know, Sal," said Jason, "I've been thinking of coming home."

"Home? Why would you do that? You're on track to be CEO of—I don't know—Prudential or something."

Bootstrappers: Forty Megs and a Stool

"I know, I know. But I've been thinking that a much bigger opportunity might lie closer to home."

"You don't want to work for some kick-ass company like Microsoft?"

"I want to do something on my own—that's what Bill Gates did. Hell, that's what everybody's doing. See, we're in a whole new era now—an entrepreneurial age. The Internet's changing everything—the future belongs to the swift, not to the strong."

"Well," said Sally, "Northborough isn't exactly Silicon Valley or Route 128."

"That's where you're wrong," said Jason. "An Internet start-up can be anywhere."

"But why here?"

"Because the opportunities are wide open in this town. Northborough's population is young, literate, and just dying to jump on the Net. But it's still a tremendous hassle to get online here. The only 'local' ISP isn't really local—it's more than fifteen miles away—that's a long distance call. Sure, you can dial in to AOL, but their throughput is a joke—I can barely get 8 kilobits per second on my 56K modem. Also, there are a lot of businesses in this town that need Web sites, and they're not going to use AOL to build them. Hell, AOL doesn't even answer their frigging phone."

"I see your point, Jason, but where are you going to get the money?"

"Remember Alex?"

"Your crazy roommate?"

"Yeah. His dad's an orthopedic surgeon, up in Killington. Pulls in almost a million dollars a year treating ski injuries. He and a bunch of his doctor pals have a little investment club—and they love this ISP idea. I'm working on a business plan, and if everything goes as I think it will, they'll front me the money. Even if we blow it, they'll just write it off for taxes—that's the way these little investment clubs work."

"It sounds fishy."
"It isn't. This is the real deal."

Wide Open Town

Jason went back to Boston the next day, and Sally returned to her low-key job at the art gallery. Soon, they began trading e-mail, and he sent her a list of books she should read. They included a host of inspirational tomes exhorting the unlimited upside financial potential of the Internet. He e-mailed her a copy of his business plan; it ran for more than seventy pages.

The plan hammered at one main point: the fact that while it was certainly true that AOL, the Microsoft Network, AT&T, and other large Internet Service Providers were likely to dominate the market for the foreseeable future, plenty of room remained for local ISPs—especially in rural markets that the big companies didn't think were important—to stake out terrain and hold it for many years to come.

AOL—the eight-hundred-pound gorilla of the dial-up market—didn't even offer local number access in more than 50 percent of the American market. Nor did other big ISPs such as Earthlink or those run by the Baby Bells. For anyone living in a rural area, getting an Internet connection remained a frustrating, expensive chore. Unless somebody did something to change this situation, most of America—which in 1997 wasn't even half-wired—would never log on to the Net, and that, in Jason's words, would be "a national tragedy." Furthermore, what business in its right mind would trust its networking needs to the likes of AOL or another giant company located hundreds of miles away? Business customers would clearly be happier dealing with a local company—one that, in Jason's words, would be "as accountable as the local General Store."

Jason's plan for CompleteNet—as he named it—was predicated on four revenue streams that would keep the company afloat. They were:

1. Dial-up accounts sold directly to the public for $9.95 a month

2. Business hosting accounts (with personalized e-mail) sold for $29.95 a month

3. Value-added services, including Web site design and hosting (pricing TBA)

4. High-speed Internet Service (when available)

Sally read through Jason's plan, and became convinced all over again that her brother was a genius. It was nearly Christmas before she heard from him again, but when he called, he was full of fire.

"Sally," he said. "I've done it. I've got the money."

"My gosh—that's great."

"So do you want to work for me?"

"Well, of course. But what would I do?"

"Well—everything. I mean, everything that I can't do."

"Jason—I almost hate to ask this, but what are you going to pay me?"

"We all have to work cheap—at least at the beginning. How's $300 a week sound?"

"Well, it's more than I'm making at the gallery. Count me in."

Modems on Main Street

Sally quit her job at the art gallery to work full-time on Jason's start-up. Her first assignment—finding suitable space for CompleteNet—was a resounding success. She found an eight-hundred-square-foot office on the floor above the only computer store in town. The rent was only $650 a month, and the space was already wired for high-speed T1 access.

"Excellent," said Jason, measuring out the floor with a tape measure. "We'll just build one wall here with a door in it, to create a front office and a back. All the ugly stuff—wires, modem racks, server boxes—goes in the back. We keep the front space clean—immaculate—for the business clients."

The division of labor was simple: Jason—with his charismatic Ivy League manner—would specialize in signing up business customers. Sally's job was, as Jason flippantly put it, to do "everything else." This included supervising the purchase and installation of CompleteNet's hardware. Sally had enough common sense to know that buying and configuring this equipment was beyond her, so she recommended that Jason pay Murray—the computer store's manager—$200 to come up with an equipment list.

"Two hundred dollars for an equipment list?" Jason complained. "Does this idiot think I'm made out of money?"

"Jason—I don't know a WAN Uplink Router from a Remote Access Server. Murray does."

"This is Info Superhighway Robbery!"

"Look, Murray can buy all of this stuff dirt cheap. We'll recover the costs on the purchase. He'll wire this stuff up for us, and he's downstairs if anything goes wrong."

"This stuff is supposed to be plug-and-play," said Jason, "but I guess we don't have a choice."

Murray was a talkative older fellow who showed Sally how all the modems and servers plugged together. While he crouched at the back of the rack, Sally peppered him with questions.

"Murray, why didn't you guys get into the ISP business?"

"I almost did," said Murray. "But AOL and Microsoft are going to own the world soon. I've seen this a million times—big companies always win—the little guys just wind up with arrows in their backs."

"Well, how do you make any money?"

"Because there's no CompUSA within fifty miles—at least right now there isn't. But there will be—someday—and then I'll be toast."

"Yeah, but your service is so much better. Doesn't that make a difference?"

"Customers are all fucks," Murray said. "They'd stab you in the back for a 10 percent discount."

"That's pretty depressing."

"That's just the way this business is," he said, crimping a set of RJ-45 cables.

The Customer Is Always Irate

Despite Murray's fatalism, CompleteNet began attracting new business within minutes of opening in February 1998. Jason's glad-handing activities had landed the company a handful of business accounts, and Sally's grassroots marketing efforts, which included designing, printing, and posting flyers all over town, had attracted more than a hundred individual dial-up users, each of whom paid $9.95 a month for unlimited Net access.

The revenue was marginal—less than $1,200 a month—but it was a strong start. People—plain folks who wanted to jump on the Net and local businesses that wanted to build their own little storefronts—were plainly looking for a way to "Put The E in E-Business" (as Sally's marketing copy had put it), and CompleteNet was the only game in town.

In March, increasing demand for CompleteNet's services caused Jason to face the inevitable—that he'd need more people to keep up with all the new business they were getting. So he directed Sally to send out an open call for two positions—a full-time ColdFusion programmer and a full-time graphics designer. Within a week, more than thirty people had applied for the positions.

Jason had his pick of Northborough's most talented and dedicated young people. But, to Sally's dismay, he only seemed interested in hiring applicants willing to work the longest hours for the lowest salary.

"We're a start-up," Jason would say to the candidate. "We don't have benefits, we don't have stock options, and we don't have any frills. Do you see any foosball tables here?"

"What's a foosball table?" asked the candidate nervously.

"Forget it. What did your last job pay you?"

"$18.75 an hour."

"How about the job before that?"

"$8.00 an hour. But that was at Kmart."

"We'll offer you $9.00 an hour. Take it or leave it."

Sally had never really seen this side of her brother before—the fact that when push came to shove, he was a complete cheapskate. Sally was sort of cheap herself. She liked living in sparse, raw spaces reminiscent of a nineteenth-century Paris garret, and didn't find it unusual that the only furniture that Jason had let her buy for the office was a set of battered chairs and tables acquired from the local Salvation Army store.

But she also wondered whether he wasn't being penny-wise and pound-foolish when it came to hiring people. Still, she was happy with the people that Jason finally hired—a programmer named Kevin and a Web designer named Maryam. They were earnest people who worked like Trojans.

By the summer, CompleteNet had twenty-seven businesses paying $29.95 a month and 260 subscribers paying $9.95 a month. While this growth was great for CompleteNet's bottom line, it was terrible for Sally, who was caught in a bind. More

Bootstrappers: Forty Megs and a Stool

business meant more tech-support calls, which seemed to come in every ten minutes or so. Sally didn't want her tiny staff diverted from their work by asinine questions about Windows Dial-up Dialogue boxes, so she wound up fielding most of the calls herself.

"Hello, CompleteNet. Yes sir. You can't log onto CNN.com? Did you type in the URL correctly? Hello, CompleteNet. Your RealAudio isn't working? Did you turn on your speakers? Hello, CompleteNet. You've got an Error 458423 Exception Error in Module CCDDDC? Ummm . . . "

Sally would always try to remain courteous—these people were, after all, her neighbors—but she also knew that she rarely was able to help them with serious technical problems, such as the infamous Windows Blue Screen of Death.

"I'm sorry, sir. I don't think I can help you."

"You can't? Well then refund my money. Return my computer to normal or I'll sue your ass in court."

"Jason," said Sally, after a long day of guiding disconnected users through their Control Panels, "I'm going nuts. These people are asking me questions that are way over my head. I know something about this stuff, but half the time I feel like it's the blind leading the blind."

"We're not making enough money to hire a full-time support person," Jason said, tightening his tie in preparation for another sales call.

"Can we outsource our support to somebody?"

"Outsource? And pay somebody else a buck fifty a call? C'mon, Sal—these idiots are lucky we're even taking their business. Let them go back to AOL if they're unhappy."

"Maybe we should hang a big sign outside the building that says "CompleteNet: We Suck Less Than AOL," said Sally wearily.

"That's not funny," said Jason, banging down the stairs to the street.

Churn

For most of the rest of the spring, Sally rarely left the office, which was a cold place once the landlord shut off the furnace on April 1.

"Well, we lost another ten customers today," Sally said to her brother as he trudged in from a sales call.

"It's not our fault—it's the goddamn upstream provider. Didn't you put the service interruption notice on our Web site?"

"People obviously can't see the Web site. They can't even get a connection."

"You're breaking my heart," said Jason, looking around the office. "Hey, where are the others?"

"They only work until 7:00 P.M."

"You mean they clock out of here on a schedule? This is a fucking start-up—they need to stay here until the work is done."

"The work *is* done," Sally said. "Kevin and Maryam built everything that's supposed to have been built. But the support calls keep coming in—that's why *I'm* still here. Why don't you send me out on some sales calls and *you* can stay here and answer some of the support calls."

"Fine with me."

"Great. Tomorrow, I'll go on some calls. You handle the phones," said Sally.

The next day, Jason stayed in the office, and Sally went on the road. She didn't have any luck convincing any merchants that they needed CompleteNet's "complete end-to-end hosting solutions," but she was happy to spend a day not worrying about IP Numbers, Default Dialers, and Line Noise. Around five, she returned to the office. Jason was at his desk, reading a book by Intel's Chairman called *Only the Paranoid Survive*.

"Hi Jason."

"Hello."

"Where are Kevin and Maryam?"

"I fired them."

"You fired them—why?"

"I didn't like their fucking attitude. Do you know what that geek asked me to do today? He asked me to buy a fucking refrigerator. A refrigerator! To stock what—beer? Do these idiots think this is some kind of frat party we're doing? This is war—business is war!"

"You're drunk, aren't you."

"I'm not drunk. Here," he said, handing a folder full of résumés to Sally. "Find some more candidates. Find some that won't talk back."

"What's happened to you?" Sally screamed as her brother pounded down the stairs. "You used to be a nice guy—not an obnoxious lunatic."

Sally found two more people to fill Kevin and Maryam's positions. They were less able, but just as willing to work for practically nothing. She did, however, vow that as long as she worked at CompleteNet, she'd never leave her brother alone with her staff again.

Burn

By the end of the summer of 1998, CompleteNet was as big as it ever would become. With 325 dial-up subscribers, 34 business accounts, and a profitable stand-alone Web design business, it was grossing almost $6,000 a month. Sally's salary had been raised to $400 a week, but she was miserable because Jason still refused to hire a full-time support person to handle the increased load.

"I just bought a 56K modem and you're giving me 8 Kbps," one user complained.

"I can send e-mail but not receive it," complained another.

"Help me, for God's sake," pleaded a third. "I've just downloaded the ILOVEYOU virus! What do I do?"

Beyond suffering the daily stress of confronting CompleteNet's users, Sally had actually begun to fear her brother, not because of any particular threats he made to her, but because of how he was now treating CompleteNet's customers.

The first target of Jason's rage was users who seemed to stay logged online for too long—thus tying up CompleteNet's modem lines. He rewrote CompleteNet's TOS (Terms of Service) to provide for immediate banishment if any users were suspected of this practice. He patrolled users' personal home pages for signs of nudity, copyrighted GIFs, or illegal MP3 Files, and wrote furious e-mail messages to the offenders that were CC'd to the FBI, the RIAA, and the Software Publishers' Association. He became an anti-SPAM crusader, and promised to sue anyone suspected of using CompleteNet's services to send even a single unsolicited e-mail message. He sent threatening e-mails to the entire subscriber base, in a tone that accused them of being in league with pornographers, illegal gambling sites, and prostitution rings.

More than anything, Sally dreaded when two support calls came in at exactly the same time. She'd pick up one call and answer it politely, but before she could hit the "Hold" button, Jason would often pick up on the other call, scream into the phone, and then slam it down in the middle of the conversation.

One evening, after the staff had left, Jason came over to Sally's desk. "Sally, I want to speak to you."

"Okay," she said. "Speak."

"Let's go over to my desk."

"Okay," she said, getting up.

"All right," Jason said, opening a folder. "Let's get down to business."

"Let's," said Sally.

"Sally—this is a performance appraisal. Not a dialogue. So just shut up and listen."

"Jason, I've been shutting up and listening for almost a year. Maybe you should shut up for awhile and listen too."

He closed the folder and folded his arms.

"All right, Sally—I'm listening."

"Do you have any idea what our customers think of us? How crummy they think our service is? Did you know that there's an entire USENET group devoted to us now called alt.completenetsucks? And it's all because you're so damned cheap that you won't hire somebody to take these calls."

"Go on, please," said Jason, his face reddening.

"The whole idea behind CompleteNet was to provide something better—something that would give people in this town a better user experience than AOL. What was it you said: 'to be as accountable as a general store'? But that's bullshit, Jason. You don't give a shit about the customers—you just want to squeeze as much money out of these people as you can by giving them the crappiest service possible. You're no better than AOL—in fact, you're worse, because AOL doesn't even pretend to care. It's hypocrisy—total hypocrisy."

"Are you finished?" Jason asked.

"I suppose," said Sally.

"Well I've got two things to say," said Jason. "First, we're a start-up. That means no frills. No luxuries. No crap. No excuses. Do you think I'm profiting mightily from this? Have you noticed an extra car in my driveway? Or any meals that I've ever charged to the expense account? Or any other luxuries that I've lavished on myself?"

"No," said Sally.

"Alex and his family have given us one chance—not two, not three—to make this business profitable, and I'm not going to let them down. You talk about customer loyalty as if it's some kind of holy pact. Well, if people aren't happy with our service, they can go somewhere else—back to AOL, or AT&T, or elsewhere."

"What was the second thing you wanted to say," asked Sally.

"You're fired," said Jason.

"Fired?" said Sally. "Are you out of your mind?"

"I've decided to hire a full-time support person to replace you," said Jason. "Somebody who knows their shit and doesn't talk back. Don't let it be said that I never listen to you."

Epilogue

Sally received a small severance from Jason—two weeks' salary and permission to use the office to find another job. Within a few weeks, she'd found one at one of CompleteNet's clients, an aerospace company that needed a full-time support specialist who could double as a Web site project manager and client liaison. During the winter of 1999, she tried to put the experience behind her, but couldn't help watching as CompleteNet's subscription base began to fall apart. The Tech Support Guru that Jason had hired was clearly doing a bad job, and users were defecting in droves. Accelerating this trend was the appearance of a new generation of ISPs such as NetZero and BlueLight. They charged absolutely nothing for a dial-up account, and when users called, they were treated like human beings—not pests.

With its dial-up business in collapse, Jason responded by repositioning CompleteNet as a B2B Web hosting company and, once again, fired the whole staff in order to, as he put it, "clean out the dead wood." This strategy seemed to work, but by the middle of the year, Jason was forced to fire himself, when his silent partner's father's investment club bailed out. The enterprise wasn't making enough money to make the millionaire doctors much richer, and it wasn't losing enough to provide a significant tax shelter either.

Bootstrappers: Forty Megs and a Stool

By mid-1999, Sally had saved several thousand dollars from her job at the aerospace company. In August, she bought CompleteNet's assets, including subscriber list, hardware, and various domain names acquired by the company—at auction for $6,000—less than half of what it was worth. She paid Murray $500 to move the equipment into the basement of her mother's house and hook it all up. After a long weekend wherein the company went dark completely, CompleteNet came back online. It was a shadow of its former self—just thirty-five subscribers paying $5.95 a month, but soon it was paying for itself, while costing almost nothing to run.

Jason and Sally have not spoken since the day he fired her and, as we write this, he's missed two Christmases and two Thanksgivings. Rumor has it that he's left the country entirely and is currently residing in Prague, where he's the CEO of a small Web shop whose main source of clients come from eLance, eGuru.com, and other bid-for-project sites. Jason's new company doesn't have much of a reputation for quality, but its prices—in any currency—can't be beat.

— Grave Robbers: Picking the Bones Clean —

Grave Robbers: Who Are They?

Grave Robbers are the necrophilic attorneys, liquidators, and investors of all size and type that profited from dead or dying dot.coms. Grave Robbers stripped everything imaginable from the New Economy's innumerable corpses: from inventory and computer equipment to slightly kicked-in cubicles, Aeron chairs, and light bulbs. In some instances, Grave Robbers dealt in more abstract assets, such as purchasing domain names or offering last-minute capital to companies that were about to go under.

Depending on who you are, and how much money you lost in the dot.com bust, you either lovingly extol Grave Robbers as twenty-first-century Robin Hoods or you hatefully dismiss them as Vultures and Bottom-Feeders. There is certainly a case to be made for both points of view.

On the positive side, Grave Robbers unlocked value for consumers and creditors alike. By buying and reselling goods that would've otherwise been shipped back to manufacturers, Grave Robbers allowed creditors to see some return on their investment and they helped consumers to take advantage of really great bargains, sometimes 70 percent off retail price. We're not talking knockoffs here either. We're talking name-brand jewelry, electronics, computers, TVs, you-name-it, which, if flooded into the market in bulk, would devalue such items to the point that manufacturers would be throwing themselves from the nearest window.

There is, of course, nothing new about profiting from someone else's financial misfortunes or "passing the savings along to you." During the Depression, speculators snapped up the assets of bankrupt railroads for pennies on the dollar; and in the late eighties, vulture firms made a killing on companies like Bloomingdale's and 7-Eleven when their junk bond-funded leveraged buyouts turned out to be as foolish as any dot.com IPO. What's more, the retail sector has traditionally turned to "jobbers" to offload excess inventory or manage distress sales by purchasing large blocks of goods and in turn selling them off in smaller blocks to wholesalers.

So why do so many people look down on Grave Robbers? Well, the reason is simple. Just as the Web has created a new, real-time distribution channel for liquidators, it has also given them a global communications network to pursue their business and, in many instances, reveal the joy they displayed in making a big "score." One need look no further than Patrick Byrne, the CEO of Overstock.com, for evidence of such glee.

Grave Robbers: Picking the Bones Clean

On numerous occasions in the media, and on postings on FuckedCompany.com, Byrne has spoken as if he's really getting his jollies seeing the mavens of the New Economy brought low. We won't go into Byrne's high jinks right now, because we devote the rest of the chapter to them. For the moment, suffice it to say that he won't be invited to speak at any venture capitalist summits anytime soon.

Whatever your feelings about Grave Robbers, they performed a necessary function during the New Economy's death spasm. They scavenged, recycled, and resold the assets that powered the glorious e-commerce dream, and paid in cash, not in empty promises. Watching them work may not be any more pleasant than watching a Turkey Buzzard eviscerate a squirrel, but predators—whether they work for Darwin or Warren Buffett—deserve your understanding, if not your empathy, because they keep the herd thin.

Are You a Grave Robber?

You might be a Grave Robber if...

- When you were a kid, you didn't run a lemonade stand. You invested your allowance in purchasing the depressed assets (namely signage, wood, sugar, lemons) of failed lemonade stands in the neighborhood.

- Major world disasters and economic downturns set the wheels in your brain turning in search of a business opportunity. (For instance, you suddenly became involved in buying lots of Persian Gulf-era American flags within twenty-four hours of the 9/11 attack.)

- You made a mint last week buying $10 million worth of teddy bears for fifty cents on the dollar, but still manage to be surly about everything and everyone, except yourself. (Whom you hold in the highest personal regard.)

- When someone calls you a "vulture," you start talking about Schumpeter's "Creative Destruction of Capitalism" and/or punch the person in the face.

- You impatiently waited for the whole dot.com thing to collapse. The only surprise to you is how long it took. (According to your calculations, it should've ended with the Asia Crisis in 1998, but then the Fed had to interfere with its bailout plan, thus postponing the inevitable moment of reckoning.)

- When times are tough for everyone else, they're great for you. (All those moon shot IPOs over the past five years must've really pissed you off, yes?)

- If you got trapped in an elevator with a dot.com CEO, you wouldn't know whether to thank him for running his company into the ground or insult him for being so out of touch with the basic rules of business. (Our prediction: you'd do both.)

Fun Facts About Grave Robbers

How They View Themselves: The last sane bastions of the free market economy in a world gone mad. (Yawn.)

How Other People View Them: The bear market equivalent of the luckiest sons of bitches on earth, vultures, scum-sucking maggots, etc. (We already talked about this, but it's just that we can't get over how much money these people are making while the rest of us are praying to find a job shoveling shit.)

Post-Bust Stress Rating (PBSR): −3 billion (which, incidentally, is also how much debt Amazon has).

Favorite Movies: The Poseidon Adventure, Airport, Silence of the Lambs, and, of course, *Jaws.*

Favorite Songs: "Eve of Destruction," "Another One Bites the Dust," "Angel of Death," anything by ABBA. (There's nothing like chilling out to some Nordic power-pop after a long, hard day of eviscerating a fallen dot.com.)

What They Do for Fun: Say nasty things about venture capitalists and, when they get tired of that, they say nasty things about twenty-something dot.com CEOs. (That's fun, right?)

People They Admire: Warren Buffett, Benjamin Graham (the father of "value investing"), Hannibal Lecter, and the Professor from *Gilligan's Island.* (After all, he was the ultimate scavenger who used coconuts to create the prototype for ARPANET. Take that, Al Gore!)

Psychological Profile: Your Assets + Your Ass = Their Profit (You got a problem with that?)

Grave Robbers: The Story of Byrne

"Would you care for water? Or perhaps some coffee?" the Prada-suited receptionist asked.

"No, we're fine, thank you. We just had breakfast," Patrick Byrne responded. "Do you think it'll be long?"

"They're finishing up another meeting. I'm sure it won't be more than five or ten minutes."

Byrne and his CFO took seats across from one another on matching leather sofas in the waiting area of Birch

Capital, one of Sand Hill's most prestigious venture capital firms. Between them, a long coffee table displayed the latest issues of such popular technology magazines as *The Industry Standard, Red Herring, Upside,* and *Fast Company*, along with a few odd white papers on the future of broadband, wireless networking penetration, and the spending patterns of women online.

"So what are the odds these people will give us $30 million?" asked the CFO, eyeing the pile.

"Nil" said Byrne. "We'd have much better odds if we were two kids right out of Stanford with a half-baked idea."

"Look, Pat, I know you're down on venture capitalists, but there are some really smart ones out there, people like you who look at companies as long-term propositions, not just investment scams."

Byrne said nothing. He stared off into the middle distance, apparently lost in thought. In fact, his ears were straining to hear words filtering through the half-open door of the conference room at the end of the hallway. This was the "pitch room," where the currently installed crew of dot.com entrepreneurs was making its big play for the big bucks. Although Byrne's hearing was keen, the most he could make out were a few disjointed snippets of conversation:

"Click-and-mortar . . . "

"Ideavirus . . . "

"Brand equity . . . "

"Exponential upside potential . . ."

"The Amazon of the online gardening space . . . "

Byrne cursed under his breath. The buzzwords were nothing but puffs of hot air—a kind of fuzzy-minded shorthand that had lately come into vogue. A few years ago, before the investment world had turned topsy-turvy, such blathering would have gotten the speaker laughed out of any self-respecting VC's pitch room. But today they were the lingua franca of the New Economy and you couldn't get money from Sand Hill without using a boatload of them.

"Do you think that if we called Overstock.com a 'Vortal' we'd have a better shot?"

"They're ready for you now, Mr. Byrne," said the receptionist, ushering them into the same conference room where the previous session had taken place.

Asleep at the Pitch

Byrne and his CFO positioned themselves at the head of a long table in front of a large projection screen. During the customary three-minute period in which PowerPoint was activated, notebook cables were connected, and color printouts were distributed, Byrne took a "read" on what they were up against.

Byrne and his CFO had been pitching Overstock.com for two weeks now—the same kind of grueling two-man road show that was a mandatory rite of passage for budding dot.com entrepreneurs seeking to take their projects from incubation stage to full-blown businesses. So far, they had had no luck attracting any interest.

In fact, Byrne and his CFO had been rejected so many times that they'd had a chance to perfect an unerring prediction system for the level of rejection they'd receive, ranging from tepid indifference to outright hostility. The system was based on studying the ages of the people in the room, how they were dressed, and the number of electronic devices that were on the table in front of them. To communicate these first-blush impressions between them, a set of hand signals was employed. If either of them scratched his nose, the signal meant "these folks might have a clue—drive in." If one of them nonchalantly tugged an earlobe, it meant "keep it simple; we're dealing with children here—let's get out of here before we're insulted."

Byrne and the CFO knew that the Sand Hill Road VCs might think that their idea for Overstock—a Web-based discount outlet—wouldn't be trendy; in fact, it's approach was so low-rent that it might earn them much scorn from these well-tailored young people. But he never appreciated that its core value proposition—using a cheap medium to distribute information about cheap goods—might be so summarily ignored. But neither of them were prepared for the amount of ear-pulling they'd have to endure—or the sense of beating their heads against the stony walls of Sand Hill indifference. Byrne refused to give up. His idea was good—not exactly orthodox—more about $20 down comforters than about $400 stock valuations. But he knew his site was popular (more than 1 million visitors per week), cheap to run, and capable of living out Byrne's basic investment philosophy, inherited from Warren Buffett's teaching, that there could be no greater command than that of "buying a dollar's worth of value for thirty cents." The site was also certainly head and shoulders above the fluff that was then in play: the preposterous "vortals," the pyramid marketing gussied up in a glossy GUI, the sticky "teen communities" and game-based affiliate networks.

Byrne took a deep breath, put a convincing smile on his face, and prepared for the opening words of his presentation. When satisfied that a quorum was present, he launched himself into his pitch.

"Overstock is a very simple idea that's as old as retailing itself. From the beginning, manufacturers have been looking for a better way to liquidate goods that are 'oversold'—i.e., closed-out merchandise. In the old days, 'rack jobbers' performed this function; they bought up the overstock and retailed the lots to consumers. But the process was both expensive and inefficient. We use the Web to let consumers instantly know about the deepest discounts on brand-name, closed-out merchandise. Our technology lets them inspect a lot, see how many items are in stock, and place an order. It's a win for both consumers and for manufacturers. Our users can get a $100 designer dress—new in the box—for $50, or a $1,200 computer for $500. With your money, we can buy a new, fully-automated warehouse in Utah."

One VC looked at his watch. Another tapped his index finger. Soon, a young man in a blue Oxford shirt, who, at twenty-eight, seemed to be the senior member of the team, asked them to "please speed it up." When it came time for the Q&A, it was plain that nobody saw that Overstock had any positive financial potential.

"B2C is dead. Tell us about your enterprise strategy," said a ruby-haired young man who looked like he was about to fall asleep.

"'Overstock'—I don't like that name," said another.

"I don't like it either" said the senior member of the team. "It reminds me of 'Livestock.' Have you considered going with something funkier, like 'eMercanto'?"

Byrne and his CFO calmly addressed these and several other concerns. Inside, though, they were both quietly fuming. "B2C is dead"—that ended the discussion. By condemning the entire e-tailing segment to the junk heap, the VC had pronounced its death.

Twenty minutes later, after being ushered out to the parking lot, the CFO exploded in a fit of uncharacteristic rage.

"That 'B2C is dead' stuff—do you know where they got it? Read this!" he said, jamming that morning's edition of the *Wall Street Journal* under Byrne's nose.

"Those people don't know their ass from a hole in the ground," said Byrne.

He put the car in gear and took the nearest on-ramp onto the crowded San Diego freeway. He'd been turned down by every VC on Sand Hill, and felt as if Overstock's prospects were as jammed up as the highway—as if the venture capitalists were only letting certain kinds of cars into the passing lane—if you weren't a late-model red sports car or a fully-loaded SUV, you could forget it. You might as well be a rusted-out, 72 Pinto, leaking gas.

"It's 1999," said Byrne, "and I think I'm the only guy in Silicon Valley who can't raise capital!"

Fundamental Stupidity

Byrne found the situation especially ironic because, unlike the twenty-something VCs that had just kicked him up and down Sand Hill Road, or the buzzword-spewing entrepreneurs who were their darlings, he actually had some real experience running companies. In Byrne's case, the companies he ran—$100 million enterprises that had performed successfully over the years—were for none other than Warren Buffett.

Buffett was both the second richest man in the world and the foremost exponent of "value investing," an investing philosophy developed in the 1930s whose main tenet is that "bad things happen to good companies." By this is meant the idea that one of the best ways to get wealthy is to seek out good companies—those with sound financial fundamentals—that have gotten themselves into a jam. By investing in them, guiding them, and otherwise helping them out of their mess, investors made money.

Value Investing had made many billionaires besides Buffett. But by the mid-1990s, just when Byrne was coming into his own as a Buffett-style VC, it fell from favor. As the dot.com bubble inflated, investors began turning away from the whole Value Investing model, not because it didn't work, but because it took too long. Identifying good companies and nursing them back to health often took months—sometimes years.

By 1997 or so, the spotlight had passed to a new generation of investors who sought to get rich by participating in "hypergrowth technology plays." Spurred on by the same magazines that Byrne had seen on the VC's coffee table, each of which announced the monthly minting of a new dot.com gazillionaire, investors weren't thinking about value anymore. Instead, they were all asking the same question: "Why invest my money in a distressed 'good company' that might take years to be healed, when I can triple my portfolio in a couple of months by finding the next Amazon or Juniper Networks?"

Byrne still believed in the immutable rules of business, but with twenty-three-year-olds walking around with 40 million dollars in play money, and people calling Warren Buffett "a dinosaur," he became convinced that the whole investment world had gone topsy-turvy. Nobody seemed to care about "fundamentals" anymore—it was a word that had become a dead term in the age of hypergrowth stocks.

Yourcoffin.Com

After having his business plan so humiliatingly rejected by the VCs on Sand Hill Road, Byrne licked his wounds for awhile. Sure, he was pissed off, but he seemed content to keep Overstock.com—which at the time was a small, ugly-looking Web site connected to a makeshift warehouse facility in Salt Lake City—running on funds supplied from his own pocket.

As 2000 dawned, he had given up any prospect of ever hearing from any Silicon Valley VC again, but then, quite unexpectedly, he started receiving calls from them. Overstock.com, for some unknown reason, was "in play" again.

"I thought you wanted to have nothing to do with me," said Byrne. "Didn't you guys just tell me that B2C was dead?"

"We're rethinking things," said the VC.

The peculiar circumstances which led to Byrne's plan being reappraised had nothing to with the merits of Overstock. Instead, they were caused by a strange, somewhat occult misfiring of the financial pipeline supplying new Internet companies with fresh VC money. This pipeline, which once had provided a robust flow of ready cash, had halted as Y2K fears mounted, but it opened up again in early 2000—for the last time.

Suddenly, business plans—especially sensible ones—became scarcer than hens' teeth. The result was a stream of completely half-baked concept-plays that marked the dot.com boom's final absurd flourish. They included such "winners" as MedicineOnline's BidForSurgery.com, YourCoffin.com, and DigitalConvergence's CueCat, a kitty-shaped gizmo that scanned URLs from magazines.

In the shadow of these malformed, misconceived enterprises, which even the most optimistic industry observers agreed didn't have a prayer, the idea of a "sensible" business plan that was not completely based on fluff or gimmickry no longer seemed as silly as it had in 1999.

By February, Byrne had received a flurry of offers. Soon, a bunch of FedEx packages containing "term sheets"—documents that tell an entrepreneur how much a VC is willing to invest, and how much he wants in return—were piled up in his office.

With his CFO at his side, Byrne sifted through the stack of term sheets, each of which outlined the sort of Faustian bargain he'd have to make with the Venture Capitalists.

"Look at this," said Byrne. "It's an insult to capitalism. They're offering us $20 million at a $100 million valuation. If we don't triple their money within a year, the $20 million comes out of our hide. Otherwise they reserve the option to approve how we run the company and at the end of three years, if they don't like the job we're doing, they get all their money back, plus interest."

"Info-Highway Robbery," the CFO huffed. "But it's still the best offer we've got."

They sorted laboriously through the term sheets, and had soon whittled down the candidates to a short-list of investors.

One of them—a consortium—seemed like the least of all the evils. In a long series of conference calls, details were ironed out. Byrne would get $30 million but basically promise the VCs back almost $100 million over three years. If Overstock.com failed, he'd be liable for the $30 million they'd fronted him.

By mid-April of 2000, the process appeared to be drawing to a conclusion. Byrne and his CFO were all ready to close on the final details of the first-round financing when the dot.com bubble finally burst, sending the entire market into the toilet, and the VC community into a tizzy.

"They're going to bail," Byrne said to his CFO.

"Have they called yet?"

"No, but they will—unless the market gains 300 points by tomorrow."

When the slide proved to be more than just profit-taking, or a minor bump in the road, the VCs called en masse, one right after another, offering variations of the same excuse for not moving forward with the placement:

"We have issues with your capacity," said one.

"Your footage doesn't scale with your business objectives," said another.

"We don't concur with your choice of Warehouse Management Software," said a third.

"Pat," said a fourth, "I'm sorry, but I can't move forward. I lost my damn shirt this week."

"I understand," said Byrne. "And I want to thank you for being the only one who didn't give me a bullshit reason for pulling out."

First Bite

Rejected for the second time by the Sand Hill VCs, but undeterred from his quest to launch Overstock, Byrne worked through the spring of 2000 doing what he'd already been doing since acquiring the company in 1999: driving to dusty warehouses, trolling for surplus goods in bargain bins, arranging for their sale, and trying to streamline Overstock.com's operations in Salt Lake City.

The site was already popular with bottom-feeders who were used to frequenting eBay for used and abused merchandise, but Byrne realized that he needed to improve his own offerings by offering wider selections of major brand-name items. With the prospect of any help from outside now behind him, he concentrated on the task at hand—getting Overstock stocked with enough surplus goods to make it a hit with consumers.

In July, Byrne was at a toy industry trade show in Las Vegas, trying to hunt down heavily discounted lots of toys to send back to Salt Lake City in time for the Christmas selling season. Browsing among the aisles of a gigantic hotel ballroom, he came across a lengthy list of toys that had all been marked down more than 50 percent.

"Manny, what's the deal with this list?" he asked the toy liquidator.

"They're all from some dot.com that went belly up. ToyTime.com"

"This list is great," said Byrne. "Is it all new-in-box?"

"Brand-new—right out of Mattel's factory."

"Do me a favor, Manny."

"Sure."

"Don't show that list to anybody else. And tell me where all this stuff is located."

Byrne stepped away from the booth and called the CFO in Salt Lake City.

"I need you to get on a plane and meet me in Wilmington, Ohio, as soon as possible."

"Wilmington, where?"

Grave Robbers: Picking the Bones Clean

"Near Columbus. Bring somebody who knows toys and can evaluate a big lot."

The next day, Byrne, the CFO, and a small team of odd-lot assessors that later became known as the "Overstock SWAT Team" were prowling around a brand-new, thoroughly abandoned distribution center in the flatlands of Ohio. They stared amazedly at the thousands of toys stretching sixty feet in the air—brand-new items from Fisher-Price, Hasbro, Lego, and Mattel—suspended on automated racks above an intricately snaking automated conveyor belt.

"I've never seen so many toys," said the CFO.

"Yeah," said Byrne. "But look at this conveyor belt. Now that's a toy we could really have fun with. Do you know what it would cost us to set something like this up in Salt Lake City?"

"Nice," said the CFO. "You should see the forklifts—not a scratch on them."

"Tell the liquidator I'll pay $3.7 million for the toys. I also want the conveyor belt, the fork lifts, the bubble-pack machines, the Warehouse Management System—the works."

They bought everything, not just the toys, but the racks, the shelves, the computers, even the toilet paper in the stalls and the light bulbs in the sockets. For about $5 million, they got an $11 million inventory of toys, along with enough warehouse gear to assemble a $2.5 million, state-of-the art facility in Utah.

Feasting

It was a great score. Byrne had stumbled across the same bonanza of liquidation possibilities that many others were taking advantage of. By the summer of 2000, when it became clear that the NASDAQ was not going to bounce back anytime soon, the speculative fervor turned from upside to downside. Instead of options being the desired objects, it was the assets of the companies that everyone was chasing like a pack of hungry wolves.

Some, like Bid4Assets.com, focused entirely on selling the computer equipment that once powered innumerable dot.coms; others, like GreatDomains.com, did a healthy business trading on "low mileage" domain names, often purchased at great expense a few years before. One company—UsedCubicles.com—traded in the slightly kicked-in cubicles, maple U-shaped desks, and wild cherry conference tables formerly occupied by many a blue Oxford-clad business development drone. At the same time, attorneys who once had eagerly drafted S1 filings pursuant to an IPO were now composing the suicide notes of Chapter 11 files pursuant to a NASDAQ delisting statement.

Aware of the fact that he wasn't the only vulture in the sky, Byrne accelerated his pursuit to the speed of thought, working around the clock to chase down any possible lead given to him by his equally untiring minions. Byrne's next big score came a few weeks later when he was in the Bay area on business. On the road, just after a meeting, his cell phone rang.

"Miadora's site just went down," said Overstock's jewelry buyer.

"Miadora?"

"They have a ton of brand-name jewelry. Rolex, Cartier, Bulgari—the works. Someone on the inside just told me that they're laying people off . . . farh farh farh farh . . ."

"You're breaking up," Byrne yelled as the phone fritzed out and went dead.

"Shit," he said, realizing that time was very short. If the buyer was right, the buzzards would find out about the lot within hours. But Byrne didn't even know where Miadora was located, much less whether the rumor was true.

"Hey," he called to a bystander. "Where is Miadora?"

"Madiera? Go to the liquor store, señor."

"No, no," said Byrne. "Wait a minute—is there a Kinko's around here?"

"You went right by it a mile back."

Byrne made a rapid U-turn across the center lane and circled back to Kinko's, where he sought out the nearest Web-connected PC. Sure enough, Miadora.com wasn't just "down"—it was "dead" meaning that it displayed the gravestone of every doomed e-commerce merchant who'd given up the ghost, the mournful, tearful, farewell screen:

"Miadora.com would like to thank our customers for your support and patronage. We could never have done it without you."

"Damn," said Byrne, leaping up from the computer. He got directions to San Mateo from the Kinko's clerk, and arrived there within an hour. It was 3:00 and the office complex was almost empty, except for a few people who were crying and packing up their desks.

"Where can I find the owner?" asked Byrne.

"He's gone," sobbed an ashen-faced marketing woman who seemed to be in a state of shock. "Everything's gone." And with that, she buried her face in her hands.

"You mean the jewelry's gone?"

"No" she said. "The dream, the concept, my options—my career—they're all gone."

"But the jewelry is still here, right?"

"In the back."

"I'm only interested in the jewelry. Here," he said, writing the short sentence, 'I HAVE CASH' on the back of his business card. "Find the owner, and give him this. Do you think you can do that?"

"Yes." she said, wiping her face.

"It's not the end of the world," Byrne said, as he watched the woman walk back through the halls toward the back office.

"Not for you," she said, waving the card at him. "You have cash."

Byrne walked out of the sad and tattered office complex, brooded for a moment on the business idiocy that had led to such small tragedies, and got back into his car.

The owner called him back within an hour. They made a deal—$2.5 million in cash for the whole load: refrigerator-sized cases filled with Rolex watches, Breitling necklaces, Bulgari diamonds. Within months, Byrne would sell it for $12 million.

"This is quite a haul," said the CFO the next day, after the SWAT team had landed, wandered through vaults, safes, and metal detectors, and added up the value of the collected gems.

"Maybe we ought to rent a couple of armored cars to truck this stuff back to Salt Lake."

"Too expensive" said Byrne. "Just put it in garbage bags and get it out of here on a regular truck."

"Garbage bags?"

"Take the safes, the metal detectors, and the computers too."

Cannibal Capitalist

As the summer of 2000 turned to fall, Byrne continued his rampage across the graveyard of distressed e-tailers, salvaging the frayed remains of once-promising companies, loading up their assets, and shipping them out to Salt Lake City. He bought Gear.com, a failed clothing e-tailer with a $14 million sportswear inventory; Adornis, a necklace and ring e-merchant whose safes were crammed with $5 million in gems; and BabyStripes.com, a tremendous lot of Gund stuffed animals and Beatrix Potter-branded baby clothing whose retail value was appraised at $450,000. Perhaps his strangest purchase was the 30,000 hats belonging to eHats.com: two containers' worth of cowboy hats, ladies' hats, sports helmets, Stetson hats, and ski hats. He paid $35,000 for the hats, which were worth almost a half million dollars.

Revenge was sweet—the founders of several of these companies had bested his own efforts to gain VC money less than a year before. Now Byrne was buying up the goods that their millions of dollars' worth of VC money had let them buy—for pennies on the dollar. It was almost as good as getting it from the VCs themselves.

But revenge became sweeter still when a reporter from the *Wall Street Journal*, the same publication that had so confidently pronounced "B2C dead" a year before, called him up. The reporter had heard of his exploits—the "Swat Team" that leapt out of the air at the first sign of a bankrupt dot.com, the drama of humbled, tearful dot.com CEOs begging for cash—and decided that it was a great story. Ironically, the piece, which dwelled at length on Byrne's conquest of Miadora, appeared just a day before Miadora's inventory went on sale at Overstock.com. Flocks of wealthy Wall Street matrons and wealthy Robber barons (with mistresses to take care of) read the story, and immediately placed orders for cut-rate Cartiers and Rolexes. Within a week he had made a million dollars in profit.

The *Journal* article was the first in what soon proved to be a long train of articles on Byrne, each of which played up his role as a bottom-feeding dot.com avenger who snapped up e-carcasses and picked them clean. If Jeff Bezos was the cute, cuddly guy of e-commerce, Byrne represented his opposite—the Angel of E-Death, delivering justice upon the heads of everyone who had ever bought into the IPO hype.

Byrne played up the media attention for all it was worth. He seized the opportunity to reinstate Warren Buffett as a visionary, and to skewer the fools who had turned the investment world topsy-turvy. When asked by *US News and World Report* to account for how e-commerce retailers could have possibly thought they could succeed, he suggested that it was because "they smoked too much dope."

Nor did he spare the VCs and wanna-be dot.com entrepreneurs who had trumped him back in 1999. "The way all those young kids with no experience were given VC to play with ... It was kindergarten with millions. They would've been better off if they had been forced to run a 7-Eleven for a year."

Epilogue

Whether or not you agree with Byrne's larger-than-life media image of himself, his over-the-top opinions, or his chosen profession—tracking down distressed companies and scavenging their remains—you cannot argue with his achievements. Without any help from any of the true sharks in the technology business—the VCs—he built a viable company from scratch and runs it like a lean machine far away from the hype, glamour, and hot air of Silicon Valley.

While predicting the future of any Web-based business is impossible, Overstock.com's prospects look good. Today, with little spent on marketing, the site's traffic is 7 million visitors a month. Its gross revenues are expected to hit $100 million in 2001, without incurring any increased costs, or any additional debt. Even the fact that the supply of distressed dot.coms will someday run out isn't a real problem—80 percent of its discount goods come from offline manufacturers, not failed dot.coms.

Byrne's success hasn't changed his aggressive bottom-feeding habits one iota. When he's not prowling around corporate fire sales, he's often cruising FuckedCompany.com, posting messages on the site's bulletin board under the pseudonym, "Hannibal," and offering finder's fees for inside tips. He continues to relish poking the finger of justice at some of the most powerful names in the technology business, including Jeff Bezos (who, according to Byrne, "has no chance of ever making it to profitability").

Why does Byrne keep up such shenanigans, especially since Overstock is profitable, still does most of its business from old-line manufacturers, and his point of view has been validated many times over by the market, the media, and his own very large bank account?

We can't say, except to note that the desire for revenge—for justice, and for a sweet turning of the tables—is almost as powerful a motivator as the prospect of buying a million dollars' worth of Tickle-Me-Elmos for $50,000.

— Lepers: Are We Not Men? —

Lepers: Who Are They?

The acid glop from the bursting Net Bubble scarred many, but none more so than the NetSlaves caste we call "Lepers." These are the mass of unemployed and unemployable former dot.commers staggering across the Web's Post-Apocalyptic landscape in a haze, wondering, "Where did my career go?"

While Wall Street Analysts and Venture Capitalists are far more despised, Lepers are the most shunned of technology workers because they don't fit into the industry's new bottom-line, skills-based priorities. Sure, during the Boom, there was such a demand for bodies that it was okay not to be a hardcore techie. What you didn't know, you could learn as you went along, or else you could get by as a "Generalist"—a low-level producer, site editor, project manager, or some other middle-management functionary a surly geek friend of ours once referred to as, "English majors who learned how to write e-mail."

We know this sounds harsh, but it's exactly how hiring managers are thinking these days. "Don't know PHP? Can't write PERL Mods while standing on your head and administering the VPN? Forget it, pal. You ain't working. Call us back when you've become a kernel wonk!"

The tragedy of most Lepers, unfortunately, is that even when slapped in the face by less-than-tactful potential employers and ever-dwindling bank accounts, many are unwilling to go for technical training. Why? One reason is ego: Being liberal arts-types, they would rather deal with high concepts than pedestrian, nuts-and-bolts realities. The other reason is that these former Generalists simply can't stand or understand code. Anything beyond a <P> tag afflicts them with the most intense math anxiety this side of eleventh-grade trigonometry.

What will happen to Lepers? They will have to reinvent themselves. Whether the reinvention happens now or later depends on the individual. If the people we know are any indication, Generalist illusions die hard, and therefore most Lepers will stew in their own juices in their crappy apartments for months on end, until finally coming to the uncomfortable realization that they have become the industry's Untouchable Class.

Are You a Leper?

You might be a Leper if . . .

- The people who run the local pink-slip party all know you by name.

- You've sent out five hundred résumés and have only been on two interviews (both of which turned out to be multi-level marketing schemes).

- You used to get fifteen calls a week from recruiters. These days, the only people calling are creditors.

- Of all the people at your last company, you're the only one who still hasn't found work. (Of course, this might have something to do with the fact that your job title was, "Chief Eyeball Evangelist," but we're just guessing.)

- You've been unemployed for approximately 623 Internet years, which in normal calendar time translates to ten to twelve months.

- Your days have been reduced from a frantic succession of meetings and metameetings to a series of absurd Beckettesque behaviors, revolving around your job search. (Get up and check e-mail; fax résumés; check e-mail again; log off to see if there are any messages; fax more résumés from leads in the paper; call the strange guy claiming to be a recruiter you met at the laundry the other day, and so on.)

- You know what "Beckettesque" in the preceding joke means. (You also know that Rimsky-Korsakov wasn't a Russian database administrator.)

(Note: If you answered "Yes" to more than one of the above, you are definitely a Leper. But don't feel so bad. We're Lepers, too, and there are thousands more like us out there.)

Fun Facts About Lepers

Appearance: Unlike their traditional rag-clad brethren, Lepers sport the T-shirts, baseball hats, and other discarded swag of a fallen dot.com.

Social Activities: In between pink-slip parties, Lepers are most commonly found hanging out at ex-employee "reunions," where everyone obsesses over what went wrong at the company, maligns former bosses, and attempts to one-up each other on how badly they got fucked. ("My severance package consisted of being tackled by a security guard on my way out of the men's room.")

Former Salary: $55,000–$65,000 per year (which certainly beats working for a local newspaper).

Current Salary: $405 per week, courtesy of Uncle Sam.

Post-Bubble Pain Rating (PBPR): On a scale of 1–10, they are a perfect 10.

Near-Term Employment Prospects: Driving a taxi, waiting tables, tutoring high school kids on the proper use of the semicolon.

Psychological Profile: Highly Unstable. Intense bursts of activity ("I sent out fifty résumés today.") and elation ("I'm going to get a job, I'm going to get a job!") alternating with complete lassitude ("I woke up at 4:00 P.M. yesterday.") and despair ("I swear I must have a big 'U' for 'Unemployable' branded on my forehead.").

Why Lepers Are Shunned: Because after pissing away billions of dollars on pointless ad campaigns and other crap, the industry has to go to the other extreme and show that it's "serious" by not hiring "non-essential" employees with people skills and (God forbid) creativity.

Lepers: The Story of Jan

She was lying on her back on the hardwood floor with a glass ashtray on her chest. There was a cigarette in her outstretched hand and it burned somewhere, but seemed to belong to a different world, as did her own fingers, which loosely cradled the half-burned butt that would in time begin to singe the inner surfaces of her index finger.

She had never been this tired before, even after spending three years in the hellishly frenetic, twelve-hours-a-day, "we want the site yesterday" Internet work environment. Back in those days, you'd get "ass-tired" or "blurry-eyed tired," but never like this—flat on your back, unable to move, or even respond to the most natural inclination of a smoker to flick the growing ash.

As if her extreme fatigue weren't bad enough, it was accompanied by a noxious, rotting smell emanating from the white apron she was still wearing. It consisted in equal parts of sweat, plus the rotting guacamole, mole chicken sauce, mayonnaise, and sour milk that Jan had been inadvertently sprayed with all day, ever since the first party she catered had begun promptly at 9:00 A.M.

She stared up at the cracked ceiling and thought of closing her lidded eyes. But then her roommate knelt beside her, her brow knit with concern.

"Jan, I know you've had a brutal day. But it's not always this bad—usually it's just weddings or bar mitzvahs, which are much easier. Next time I'll make sure we work one of those."

"Next time?!" Jan murmured, barely able to mouth the words. Would there, could there possibly ever be a repeat of even a single moment of this demonic, enervating day?

Garbage Day

The day had begun at 6:00 A.M. on a bench in a Brooklyn subway station. By 7:30, Jan and her roommate had made it as far as Randall's Island. Within minutes, they had crossed under the approach to the Triboro bridge and found themselves standing on a grassy area where a two-acre tent had been pitched the night before. A whistle blew and the two women, along with the other members of the WeCanCater team were summoned to an area just outside the tent's entrance. A bald, athletic man wearing an apron was standing on a white metal folding chair. He soon launched into a rapid-fire pep talk, using an electronic bullhorn.

"Can everybody hear me in the back. Okay—anybody new here?" Ten or twenty hands went up.

"All right then. For all you newcomers, and I can see there are lots of you today, I just want to say that it's going to be a long day, but it's going be fun. And I want everybody to be remember our motto. Does everybody know our motto?"

"We smile," said a few people half-heartedly, "and we serve!"

"What's that? I can't hear you."

"WE SMILE AND WE SERVE," yelled the crowd.

"That's right. Now today, there are going to be lot of people here, and if anybody gives you trouble, just keep smiling, keep serving, and keep moving. Okay—let's get to work!"

The crowd quickly moved to a white van, where they were given chefs' hats, aprons, and plastic serving gloves, along with instructions about which stations they'd be working that day.

Jan's first assignment was at the coffee and juice table. By lunchtime, she had moved on to the grill, where she flipped burgers, boiled dirty water hotdogs, and plucked insects out of the potato salad. By three, she was alternating between the cotton candy cart and the mosh pit of people lining up at the Häagen-Dazs cart.

When dinner time rolled around, her brain felt deep fried. The hot sun, the constant demands of the obnoxious, hungry crowd, and the fact that she hadn't done any real physical labor for more than five years left her barely able to stand. What kept her going was simply this, that if she didn't finish out the day, ladle up another three hundred pounds of ziti, and smile and serve until the bitter end, she wouldn't come home with a single dollar. By sheer force of will, she did—by 8:30 P.M., the party was over.

As the train pulled into Court Street, the doors opened in front of a dot.com ad. Limping slightly, Jan exited the train, feeling that her past life—in the front lines of the Net Revolution—had happened centuries ago.

Online Community Redux

Like millions of other Americans, Jan's first exposure to the online medium came in the form of a gaily-packaged AOL disk. But unlike many, who simply loaded the disk into their PCs, or threw them out, Jan became enthralled by this new medium, which let her make friends from across the world. Soon, her nightly forays into cyberspace became her major form of entertainment and community. She spent hours sharing the most intimate details of her life with complete strangers, and was soon moderating several forums on a volunteer basis.

Enter Jan

Jan, employed at a white-shoe Wall Street investment house, was aware that millions of dollars were being made—or at least spent—building the bulletin boards and chat rooms that powered the online community. But she'd never really considered it as a career option, until one night—on Echo, a New York City–based electronic salon—she heard about a job opening at "Ennerva.com." She sent off her résumé and, within a month, landed a job there at a salary of $65,000 a year—about $20,000 more than she was being paid at the investment house.

On the face of things, Jan was stepping into a great opportunity. Ennerva.com was a group-buying portal with a mission statement to become "the Amazon.com of the enterprise wholesale space" and "take out" brick-and-mortar business behemoths such as Staples, OfficeMax, Office Depot, and even Kinko's. With a battle cry to "arm small businesses everywhere with bulk-buying firepower," Ennerva joined Mercata.com, BizBuyer.com, Office.com, SmallBusiness.com, and other B2B upstarts seeking to take envelope buying into the twenty-first century.

For Jan, joining Ennerva.com was simply her big chance at the brass ring that had so far eluded her. Although she was hardly the mercenary type, the money, the power, and the prestige appealed to her basest instincts. More nobly, she looked forward to the chance of building her own online community from scratch, applying everything she had learned in the last three years, and taking her own career to the next level. Also bolstering her enthusiasm were the people running Ennerva—they were all Harvard M.B.A.s whose passion to win at all costs convinced her that maybe they were all really going to take over the world.

Jan began work as community director on a Friday afternoon after accepting the job only a few hours earlier on the phone. David, the CEO, was duly impressed by her enthusiasm, and grew convinced of Jan's prowess as the day turned into night, and the night turned into morning, which found Jan still tapping away at her keyboard, speccing out the plans for Ennerva's bulletin boards and chat areas.

"Wow, you don't mess around," he remarked as he headed out at 4:30 A.M.

"You didn't hire me to sleep," she responded, without taking her eyes off the screen.

Jan didn't have to work around the clock every night at Ennerva.com, but she rarely got out of there before 10:00 P.M., and often worked Saturdays and Sundays. There was more than enough work to be done. She had approximately four months to create multiple online communities for a warehouse worth of product categories: fax machines, printers, office furniture, office supplies, and even a dedicated forum devoted to ink cartridges. Her drop-dead deadline was May 12, 2000—a number that her slightly sadistic IT manager had incorporated into everyone's screen-saver.

Although Jan had seen her share of stressful situations while working on Wall Street, she quickly found herself becoming overwhelmed by her ever-lengthening "to-do" list, which included finding bulletin board and chat software, overseeing the creation of graphics, and making sure that everything was compatible with Ennerva's buggy fulfillment system.

Also adding to her stress level was the fact that the same Harvard M.B.A.s who had so impressed her during the interviewing process refused to take her seriously during meetings. For them, it was all about marketing and sales, not about making all of the pieces fit together for the user. None of them had been online for more than a year, or had a clue about how to construct a working Web site.

Things came to a head in early January, during a Monday Morning Progress Meeting. Jan, who had worked the entire weekend, as usual, sat there bleary-eyed as the COO began chiding her for failing to meet a set of critical goals.

"According to that GANNT chart on the wall, we were due for a fully functional Alpha of the Community Area last Friday."

"It would have been there, but as I noted on the memo I sent last night, I'm still awaiting final approval on the platform, as well as a feasibility assessment on how compatible it is with our back-end system."

"Didn't we all settle this last week?" asked the Chief of Business Development.

"We only agreed," said Jan, "that we would take it up again as soon as possible."

"How about now?" asked the CTO. "Can we nail this thing down, and move on?"

"I'd love to," said the COO. "But I've got another meeting downtown in twenty minutes. How does 9:00 work for everybody?"

"You mean tomorrow morning?"

"No, tonight."

The 9:00 P.M. meeting came and went, with the only issue settled being that the team wouldn't move forward without the "opt-in" of everyone else. This further delayed matters, putting her hopelessly behind. The platform issue wasn't the only one pending approval—there were others: the user interface, the chat client, and the moderation tools. Despite her wretched state of affairs, she remained determined to meet the May 12 deadline. She had come so far, and she wasn't about to let a bunch of stuffy procrastinators make her fail.

The Day Off

By February, she had made so little progress, and was feeling so stressed out, that she called a therapist that one of her old friends on Echo had recommended as a specialist in "dot.com head cases." Although her busy schedule didn't give her much free time, she managed to visit him twice a week at lunchtime. The therapist took one look at Jan and, although he wasn't a Freudian, made her lie down on his leather couch, fearing that she was about to drop dead.

"Do you have a sleeping disorder?" he asked upon waking her.

"I don't know. I've had problems sleeping before. I guess I just have too much to do at work."

Jan recounted the litany of frustrations she had experienced since joining Ennerva and, with tears in her eyes, hysterically exclaimed that she didn't know how she could meet her May 12 deadline.

"Do your bosses realize how upset you are?" the therapist said, reaching for a box of tissues.

"They don't have the first fucking clue about what's going on. Their heads are so far up their asses that it would take an industrial strength crane to pull them out."

"It sounds to me that you could use some medication."

"I don't want any medication," Jan snapped. "I need the asshole IS guy to install the bulletin board, I need them to finally approve the design, and I need them to let my staff go in and start beta-testing the site. And I need you to stop acting like my clueless managers with these dumb questions!"

The therapist paused for several minutes, taking it all in.

"It sounds to me that the best medicine that I could prescribe for you would be a few days off."

"A day off?" she screamed. "I've got a computer with a big May 12 on it. I'm a marked woman, don't you understand?"

With that, Jan stormed out of the office, ashamed that she'd wasted $75—and worse, one whole hour—talking to such a knucklehead.

The Beating Goes On

Back at work, Jan rejoined the march toward Ennerva's soft launch—that twilight period in which the product was hyper-scrutinized to see whether it could withstand the force of millions of paying customers aching to buy office supplies in bulk.

By late March, thanks to Jan and her staff's undying persistence, almost everything was in place: The conversational areas were stocked with helpful tips, the FAQ and Help Section areas had been written and approved, the user interface had been finalized, and the back-end software seemed to be behaving itself.

Jan, however, was on the brink of a nervous breakdown. She was averaging less than two hours of sleep a night, never saw her family or friends, and even her roommate, who generally tried to stay out of her personal life, became deeply concerned about how sick she looked.

"Just leave me alone," Jan screamed, slamming the bedroom door.

In the weeks that followed, Jan's mental state deteriorated further. Once during a meeting with one of the developers to go over the Profiling Mechanism on the Bulletin Board, she passed out in mid-sentence. She was only able to be revived after several cups of water had been splashed on her pale face.

While her managers took it as a minor setback, adjusting the GANNT chart accordingly, Jan was frightened enough to call up the therapist, whom she hadn't seen in over a month. He warned her that taking a day off was no longer an option—it was "doctor's orders."

"I'll never convince them," protested Jan.

"You'll never find out if you don't try," the therapist said.

On April 12, she formally requested from the COO that he allow her to start working a six-day week.

"Look," he said, "I know you've been under the gun—we've all been under the gun." Looking at the screen saver, he reported "We're only twenty-nine days, sixteen hours, and eleven minutes from D-Day. Can't you stick it out 'til then?"

"Dan," she said, "I need the time. It's been so long since I've had any time on my own. Can't you spare me for just twenty-four hours? I'm sure you all can hold the fort."

"Jan—you're a key player here. I don't have to tell you how much we're all relying on your team to pull us through this critical period. Our company is under the microscope now. You don't get a second chance to make a first impression. Everybody—the investors, the press, the customers—they're all watching us now. If we fail, if we don't produce, if we don't go the full route—110 percent—what will the whole thing have been for?"

"You've got me."

"What?"

"I mean—I agree—we must perform."

"We must, Jan."

"So, can I have my day off?"

"I'll see what I can do," he said.

Jan worked the next seven days, flat out. The site was now on its feet—ready to embrace the throngs of customers who would soon materialize out of the ether. The conversational areas were teeming with relevant threads, the ad campaign was in place—and everything was a go.

Jan marveled at how quickly her formerly sluggish colleagues were moving, and felt proud that the product was now finally taking shape. As a tiny reward for her efforts, she had been granted not the day a week she had requested, but a single day to "do whatever she saw fit."

Jan's day off turned out to be a Saturday. Seeking to make the most of these precious twenty-four hours, she'd made plans to see some friends for dinner and maybe even to go see a movie. But because she hadn't correctly calculated how deeply sleep deprived she was, she wound up sleeping for thirty-six hours straight. Emerging from her bedroom at 10:00 A.M. on Sunday morning, Jan saw her roommate sipping coffee and leafing through the magazine section of the newspaper.

"What time is it?" Jan asked, yawning and stretching her arms.

"Time for you to go to work, dearie."

"No, I'm off today. It's Saturday."

"No, honey. It's Sunday."

Jan instantly burst into tears and crumpled to the floor into the fetal position. The more her roommate tried to calm her down, the more hysterical she became. Believing that Jan was having a psychotic episode, the roommate immediately called the therapist.

"Can she come to the phone?" the therapist asked.

"I don't think so. All she does is scream 'I can't work today'."

The therapist instructed the roommate that under no under circumstances was Jan to be allowed to leave the apartment. He prescribed a week of rest and a 250 milligram bottle of Darvon—a powerful anti-psychotic medication. He even went so far as to fax a letter to Ennerva's CEO apprising him of the situation.

The Big Slackoff

Despite the fact that Ennerva's management was aware of Jan's dire condition, they didn't let up. She was called upon to implement a whole new set of specifications—including a total overhaul of her conversational areas—that had been recommended by a focus group. Outraged at management's insensitivity, she decided to find another job, but first she was going to do something she'd never done in her three years of dot.com-mery—she was going to slack the hell off.

It came easy at first, and felt like heaven. She'd get up at noon, read the papers, watch some TV, and clean the house. When her roommate came home, Jan would make the both of them dinner. Sometimes, they'd have a few friends over. It was fun, but as time wore on, Jan found herself becoming a bit depressed; something about having so much time on her hands felt wrong, fake, like a party that went on far too long—a party that had to end.

At first, money wasn't an issue. Jan had about $25,000 worth of About.com options that she'd cashed in, ensuring that she was in no danger of missing a rent payment or a utility bill. Her biggest problem was staying busy, but she dealt with it in the way that many dot.commers do—by buying a new, fancy, multi-media equipped computer, and stocking it with games, the most interesting being Maxis's The Sims, an interactive neighborhood simulator.

She knew that sooner or later she'd have to go looking for another job. Gradually overcoming her fears about explaining what had happened at Ennerva, she began testing the waters, using job boards, news sites, and her old About.com contacts to narrow her search.

To her dismay, her quest for full-time employment yielded absolutely nothing.

"What's going on here?" she groused to a friend. "I've never had trouble finding something before."

"Don't you read the papers? The NASDAQ is in the toilet. Everybody's retrenching. Companies are going out of business left and right."

"Really? I must have really been out of it."

"Yeah. I even saw that crazy company you were working for on FuckedCompany the other day."

"Fucked Company. What's that?"

"Some kind of dot.com death site."

By June, Jan had registered with five different job sites and answered at least eleven different open positions. But nobody was calling her back. Throughout July, the silent treatment grew worse—no jobs except for unpaid intern positions appeared in her area of expertise. By August, she was running out of money, and beginning to panic.

On The Streets

For the next six months, Jan looked for work, but continued to find nothing. Like many people who had suddenly been rendered obsolete, she spent most of her time cocooning, and rarely went out, except to CompUSA to buy résumé-grade printer paper, or make a monthly pilgrimage to a local pink-slip party, where she commiserated with people in equally desperate circumstances. Although most tried to keep up a brave front, the trend was clear. The employees of yesterday's Web superstores were today's janitors, deli clerks, and bus boys.

"What the fuck," she said to her roommate one night following a pink-slip party. "Those pink-slip people are losers."

"Sounds to me like you'd better downscale your own expectations," the roommate said, before tossing a copy of the *Village Voice*—the paper that brought New York's chronically underemployed a fantastic assortment of sleazy, low-end jobs—across the kitchen table. Swallowing her pride, and biting the bullet, Jan started circling some of the ads, and later that week she began a fruitless quest for a job—any job—that she might be able to do.

The first interview she went on—an "event-marketing job"—was basically a cattle call. She showed up, but instead of being interviewed alone, she was interviewed along with four or five other people. Nobody—least of all the interviewer—seemed to have a clue about what the job was about, except that "our best event marketers make at least $50,000 a year." Jan walked out in disgust.

Her second interview was for a sales job at a local DSL provider. But Jan, who was so afraid to blow what seemed like a really good opportunity, took the wrong train to the interview, arriving three minutes late. As she raced down the hall, she used her cell phone to call the interviewer, who brusquely told her, "Sales is about timing. What if I were a client?" before slamming down the phone.

Any lingering doubts about her "Leper" status in the Internet Industry were extinguished one afternoon when she spoke to a tech recruiter. Instead of the usual, "I'll call you if I have anything that matches your qualifications," the recruiter gave her a tongue-lashing.

"You want to know why I'm going to throw your résumé in the garbage? Because all you online community people, all you editors, you liberal arts grads who got into this industry expecting an easy ride should just go away. If you can't code, hit the road!"

Jan hung up in tears. Desperate, and with no other job prospects, she broached a topic that she'd long been avoiding discussing—doing the same kind of work that her roommate had been doing—shoveling potato salad for WeCanCater.

"I don't think you're cut out for this kind of work," said the roommate. "Catering isn't anything like the dot.com industry. You don't get to sit around all day in a nice office writing e-mails. It's hard, hot, stinky work."

"I don't really have any other options right now, unless you're ready to find a new roommate."

Epilogue

Jan worked with her roommate at the catering job throughout the summer of 2001, avoiding double shifts as best she could. Just as things seemed ready to pick up, the events of September 11 sent the market into a tailspin. She applied to one job after another, sending out thousands of résumés, but got nothing in return. By March, she wound up right back where she had started—as an admin working at a financial services firm on Wall Street.

After being unemployed for so long, Jan was very happy to collect a check every two weeks, even though it was less than half of what she had collected at her technology job, and slightly less than she had originally been making. In July, her luck again ran out when the firm, plagued by allegations of fiduciary scandal involving Enron, was forced to lay off 10 percent of its staff.

As we write this, Jan is still unemployed and, like many tech workers, completely clueless about what she should do next. Depending upon the day of the week and the position of the moon, her plans for the future range wildly from work-from-home taxidermy to HVAC repair.

Afterword

— What Next? —

How are you feeling after reading these stories? Are you depressed? Anxious? Or are you in fact relieved—that you never got mixed up in this Internet nonsense in the first place?

For us, the mood is befuddlement, especially about what's going to happen next. Will things suddenly go back to the free-market, capitalist orgy of 1999? Will the tech industry be reborn—perhaps in a distant land—using the same cyber-coolies who once populated our shores? Or is the whole "IT sector" destined to become a secondary service category on par with fast food and plumbing?

We don't know. But one thing we are sure about is that we—and probably most technology workers—would take our old miserable jobs back in a second.

Ironic, isn't it, that the same people who once so openly and incessantly bitched about what they did for a living would now so willingly return to the same industry that abused them. It's not that we're gluttons for punishment, or nostalgic for "the bad old days," it's just that jobs are so scarce that even the worst one is preferable to being unemployed.

People tell us that the thing tech workers need to do is "reinvent" ourselves once again. Unfortunately, doing so is easier said than done. Dot.com experience might as well be the "Mark of Cain" upon your résumé. It confirms in the eyes of hiring managers that you are nothing more than a job-hopping, arrogant, latte-sipping, scooter-riding scoundrel who'd be out of place even in the mail room of any self-respecting American corporation.

Are we bitter? Are we angry? Not anymore. We worked through these emotions a long time ago, along with any joy we felt watching all of those stupid companies go out of business. For us, the name of the game now is survival, at any cost. No job is beneath us, and no wage is too low. We're not proud, and we no longer think that we deserve millions of dollars a year for moving pixels around on a computer monitor.

If we are, however, entitled to one last illusion, one thought to hold in our heads as we wait tables, hang Sheetrock, and avoid calls from our creditors, it is this: The Internet is a beautiful thing. And we helped build it.

Index

Adbusters, 13
Adobe Photoshop, 33
advertising campaigns, dot.com, 88–92, 151
Airport, 171
Aliens
 general characteristics of, 99–101
 H1B program, 98, 102–104
 conditions in "holding pens" of, 109–111,118–119
 story of "Dana," 101–120
Amazon.com, 46, 97, 150, 171, 172, 186, 196
America Online, 3, 51, 56, 57, 153, 154, 157, 159, 195
 takeover of Time Warner, 132–145
 in wireless industry, 56–57
 in local U.S. ISP sector, 153, 154, 159, 163
 "Greenhouse" program, 3
AT&T, 59, 154, 163
anti-trust issues at Time Warner, 134–135
AOL Time Warner, 132–145
ARPANET, 171
Asia Crisis, impact on IPO pipeline, 13, 170
Astronet, 9

B2C, demise of sector, 175, 177, 185
BabyStripes.com, 184
Baby Boomers, xi
Barlow, John Perry, 9
Barron's, 133
Barrymore, Drew, 99
Bartiromo, Maria, 148
Barth, John, 9
Bear Sterns, 57
Beckett, Samuel, 91, 192
Bezos, Jeff, 185, 186–187
Bhatia, Sabir, 97, 98
BidForSurgery.com, 178
Bids4Assets.com, 182
BizBuyer.com, 196
Blackberry, RIM, 40, 131, 141
Blodgett, Henry, viii, 47
BlueLight, 164
body shops, 112, 114
bohemianism, 11, 126
booth babes, 59, 64
Bootstrappers

 general characteristics of, 147–151
 do-it-yourself ethos of, 147–148
 rejection of IPO mania, 148–149
 gloating of, 151
 miserliness of, 147–165
 story of "Sally," 151–165
Bressler, Richard, 32
Bronson, Charles, 48
bubble, Internet (*see* NASDAQ crash)
Buffett, Warren, 169, 174, 176, 185, 189
Burning Man, 11
Bush, George (senior), xi
Business 2.0, 24
bust, Internet (*see* NASDAQ crash)
Byrne, Patrick, 168, 171–187

California, 100,
CarsDirect.com, 35, 40
Case, Steve, 135, 142
Cassady, Neal, 34
CDNow.com, 40
CEOs, dot.com, 80. 83, 128, 170, 171, 179
CFSB, 57
Chomsky, Noam, 4
CitySearch.com, 35
Clinton, Bill, ix
CNBC, viii, 40
CNN, 73
Cobain, Kurt, xi
Collins, Joe, 134
COMDEX, 64
Commodore 64, 32
Compaq, 57
Computer Associates, 57
Computer Shopper, 33
Condé Nast, 15
Cool Site of the Day, 3
Corso, Gregory, 34
"Creative Destruction of Capitalism," 170
Critical Theory, 5

day traders, viii, 71
DEN (Digital Entertainment Network), 34
Digital Convergence, 178
Disney, Walt Inc.,134–135

Dobbs, Lou, 73, 75
Dotcomscoop.com, 48
Dreamworks, 34

Earthlink, 154
Eastwood, Clint, 48
eBay.com, 180
Echo, 196
egalitarianism, of Internet, 2, 4, 9
eGuru.com, 165
ECRM, 55
eLance, 165
Eggers, David, 3
EHats.com, 184
Ellison, Larry, 83
Enron, vii, 204, 206
Enterprise Gateway Technology, 56
Entertainment Weekly, 133
Ericsson, 54
e-tailers, initial success of, 8
eToys.com, 35

Fark.com, 5
Fast Company, 74,172
Federal Communications Commission, 64, 135
Federal Trade Commission, 134, 141
Ferlinghetti, Lawrence, 20
Fisher-Price, 181
Flux, 3
food service industry, as career alternative for Lepers, 193–195, 205
foreign technology workers (*see* Aliens)
Foosball, x, 158
Foucault, Michel, 9
FrontPage, Microsoft, 43, 127
Fry's Electronics, 33
Fuckedcompany.com, 25, 47, 186, 203

Gates, Bill, 28,137
Gazoontite.com, x
Generation X, xi
Geocities, 127
Gear.com, 184
Gilligan's Island, 171
Ginsberg, Allen, 22
Goto.com, 35
Graham, Benjamin, 171
Grave Robbers
 general characteristics of, 167–171

rejection of IPO mania, 168, 170
economic fundamentalism of, 171–173
story of "Byrne," 171–187
Great Tech Gold Rush, vii–xii
GreatDomains.com, 182

H1B program, 102–104 (*see also* Aliens)
Half-Life, 47
Handspring, 54
Harinarayan, Venky, 97
Hatch, Orrin, 134
Harvard M.B.A.s, 152, 196–197
Hasbro, 181
Hecklers Online, 9
Helms, Jesse, 134
holding pens, 108–120
HotJobs.com, 128
HotWired (*see Wired*)
Howard, Ron, 34
hypergrowth technology plays, 176–177

Icebox.com, 34
iGuide, 8,
Immigration (*see* Aliens)
Immigration and Naturalization Service, 101
Independent Consultants (*see* Panhandlers)
Industry Standard, The, 24, 46, 47, 74, 172
Information Superhighway Robbery, 37, 156
Inside.com, 8
Instant Messaging, 55
Intel, 103
Internal Revenue Service, 30, 31
Internet World, 59
IPO Mania
general discussion, xi
impacted by Asia Crisis, 13
in Wireless industry, 51, 59
Internet Service Providers, 55, 148, 153–155, 161, 164
Isaacson, Walter, 136, 137, 138
ITAA, 103
IXL, 35

J.P. Morgan, 57
Jarry, Alfred, 9
Java, 61, 115
Javascript, 53
Jaws, 171
Joyce, James, 20
Junglee, 97
Juniper Networks, 177
JustBalls.com, vii

Kaplan, Philip, 25 (*see* also FuckedCompany.com)
Katzenburg, Jeffrey, 34
Kerouac, Jack, 20
Killer App, the, 53
Kinko's, 183, 196
Kissinger, Henry, 138

Lecter, Hannibal, 37, 171
Lego, 181
Lepers
general characteristics of, 189–193
shunning by IT industry, 190
exhaustion of by 24/7 pressures, 193, 202
downward mobility of, 202–205
career reinvention attempts of, 205–206
reinvention in food services industry, 193–195
story of "Jan," 193–206
Levin, Gerald, 132, 136–138
Liberal Arts majors (*see* Neo-Luddites)
Libertarianism, x, 47
Liquidators, Online
business plans for, 175
as shopping channel for Bootstrappers, 150
lobbying, for H1B program, 103, 104, 106–107
Logan, Don, 136,137
Los Angeles Basin, tech industry development in, 34–36, 103
Lycos, takeover of *Wired*, 15–16

Macromedia Flash, 33, 42, 43, 55,127
MarchFIRST, employee lockout at, 4
Mattel, 181
McCain, John, 103
McLuhan, Marshall, 9
McSweeney's, 22
Meeker, Mary, 46, 47
Melville, Herman, 20
Metcalf, Jane, 13
Mercata.com, 196
Merganthaler-Linotype, 127
Mergers
in Wireless Industry, 59
at AOL Time Warner, 132–139
Miradora.com, 182–184
Microsoft, 9, 53, 103, 154, 157
Milton, John, 24
Mirsky's Worst of the Web, 3
Mondo2000.com, 3
Monk, Thelonious, 13

Monster.com, 128
Morrison, Jim, 23
Motorola, 54, 103
Motley Fool, the, 9
MrCranky.com, 5
Mungo Park, 9

Nader, Ralph, 137
NASDAQ Crash, 4, 59, 73, 94, 179, 181, 203
National Administrative Professionals Appreciation Day, 121
NPR, 5
Neo-Luddites
general characteristics,1–5
romanticism of, 2
comp-lit background of, 9
bohemianism of, 11
career reinvention attempts of, 10
view of Internet Revolution, 7, 24, 34
rejection of IPO Mania, 7,16
downward mobility of, 21–24
story of "Charles," 5–24
Netscape, significance of IPO, ix, xi, 4
NetZero, 35, 164
New Literature, 9
New Wave, French, 5
New York Observer, The, 133
New York Times, the, 135
New Yorker, the, 22
Nirvana, xi
Nokia, 54
noosphere, 13

Office.com, 196
Office Space, 5
OfficeMax, 196
O'Reilly books, 47, 72
Omidyar, Pierre, 99
Online communities
as outlet for rage of Vigilantes, 46, 47
as gloating channel for Grave Robbers, 186
pressures of deploying, 174, 197–202
Onion, The, 3
Open Source Crusaders, 13
Oracle, 53, 103
Organic, 25, 35
Overstock.com, 150, 174–187

Packet, 3
PageMill, Adobe, 127
Palm, 54
Panhandlers
general characteristics of, 25–29

Index

rejection of IPO Mania, 29
independent ethos of, 37–39
impatience with mismanagement, 37–39
business problems of, 29–32, 41–43
story of "Caleb," 29–43
Paperclip, Mr., 19
Paradise Lost, 24
Paramount, 34
Parsons, Richard, 130, 132
Paternot, Stephan, 47
Pathfinder, 8, 136–138 (*see also* AOL Time Warner)
Pawns
general characteristics of, 121–125
attitude towards Tech Support, 124
employment humiliations of, 123–125
career reinvention attempts of, 127–130
story of "Kaye," 125–145
Peace Corps, 21
People magazine, 133
PERL, 127
Philbin, Regis, 101
PitchForkMedia.com, 5
Phone.com, 59
pink-slip parties, 191, 204
Pop.com, 34
Poseidon Adventure, The, 171
pornography industry, as a career alternative, 43
PowerPoint, 173
Proctor & Gamble, 77, 81, 85

Razorfish, 25, 35, 38
Recruiters, amorality of, 100
Red Herring, 24, 172
RIAA, 162
Rimsky-Korsakov, 192
Ross, Steve, 134
Rossetto, Louis, 13
rumor-mongering:
by Vigilantes, 46–47
at Time Warner, 137
Rushkoff, Douglas, 9

Sagan, Paul, 137
Salon.com, 8
San Francisco
bohemian allure of, 11
web design companies in, 35
exodus from, 21
Schadenfreude, x (*see also* Vigilantes)
Seagal, Steven 48
Secretaries (*see* Pawns)
Severance packages, 47, 93, 144, 192

Shape-Shifters
general characteristics of, 71–75
career reinvention attempts of, 80–82
embrace of IPO Mania, 72, 80, 87
impact of NASDAQ crash on, 73
greed distorting judgment of, 89
hard-drinking habits of, 74, 91–93
family pressures on, 87–88, 93
downward mobility of, 93–94
story of "Gene," 75–95
Sheetrock, 134, 208
Silence of the Lambs, 171
Simpson, Homer, 100
Sims, The, 203
SmallBusiness.com, 196
Snyder, Gary, 34
Software.com, 59
Software Publishers' Association, 162
Sony, 34
Space.com, 73
"Spot, The," 3
Spielberg, Steven, 34
Stamps.com, 40
Staples, 198
Stanford University, 172
Startups, 85, 122, 163
stock options, ix, 28, 48, 53, 59, 60, 65, 183
Stay-Puft Marshmellow Man, 100
Stewart, Martha, 78
Stupid Money, 21, 148
Suck.com, 3
Survival Research Laboratories, 5–6, 22
Swayze, Patrick 48
Sybase, 115

Term Sheets, for Venture Capitalists, 178
TheGlobe.com, ix,
Thestreet.com, 8
TotalNY.com, 3
TVGuide.com, 8
Twentieth Century Fox, 34

Unabomber Manifesto, 1
unemployment, 21–24, 191–192, 203–205
Upside, 172
UsedCubicles.com, 182
US News & World Report, 185
USENET, 61, 163

value investing (*see* Buffett, Warren)
Van Gogh, Vincent, 139–141

Vaporware, 53–56
Viant, 35
Vigilantes
general characteristics of, 45–49
libertarian mindset of, 46
rumor-mongering of, 46
self-identifications of, 48
greed distorting judgment of, 53, 56
story of Vincent, 49-69
Visor, 54
Vodafone, 59
Venture Capitalists, ix, x, 40, 178–179, 185, 186, 189
Village Voice, 204
Vydec, 127

Wall Street Analysts, 45, 189
Wall Street Journal, 57, 80, 133, 139, 175, 185
Wall Street Journal Online, 9
Wang, 127
Warehouse Management Software, 179, 181
Warner Bros, 13, 36, 130 (*see also* AOL Time Warner)
Wayne, John, 48
Web content failures, 3, 8
Web design companies
predatory business practices of, 36–38
bad management and staff misbehavior at, 38
WELL, the, 10
Willis, Bruce, 48
Windows Blue Screen of Death, 159
Wired, 15, 16, 74, 137
as beacon for futurism, 9–10
founding by Rossetto and Metcalf, 13
takeover by Condé Nast, 14
takeover by Lycos, 15
discrimination in cafeteria of, 14
story of Charles, 5–24
Wireless industry, 53–54
merger-mania in, 59
AOL activity in, 56
Word.com, 3
WordPerfect, 127
WordStar, 127

XML, 55, 61

Y2K
remediation projects, 116
impact on IPO pipeline, 178
Yahoo!, 20, 23
Yoo-hoo, 23
YourCoffin.com, 178

Books from Allworth Press

The Soul of the New Consumer: The Attitudes, Behaviors, and Preferences of E-Customers
by Laurie Windham with Ken Orton (hardcover, 6½ x 9½, 320 pages, $24.95)

Dead Ahead: The Web Dilemma and the New Rules of Business
by Laurie Windham with Jon Samsel (hardcover, 6½ x 9½, 256 pages, $24.95)

Emotional Branding: The New Paradigm for Connecting Brands to People
by Marc Gobé (hardcover, 6½ x 9½, 352 pages)

Citizen Brand: 10 Commandments for Transforming Brands in a Consumer Society
by Marc Gobé (hardcover, 5½ x 8½ pages, $24.95)

The Entrepreneurial Age: Awakening the Spirit of Enterprise in People, Companies, and Countries
by Larry C. Farrell (hardcover, 6½ x 9½, 352 pages, $24.95)

Turn Your Ideas or Invention into Millions
by Don Kracke (paperback, 6 x 9, 224 pages, $18.95)

Career Solutions for Creative People
by Dr. Ronda Ormont (paperback, 320 pages, 6 x 9, $19.95)

The Money Mentor: A Tale of Finding Financial Freedom
by Tad Crawford (paperback, 6 x 9, 272 pages, $14.95)

What Money Really Means
by Thomas K. Kostigen (paperback, 6 x 9, 240 pages, $19.95)

Writing for Interactive Media–The Complete Guide
by Jon Samsel and Darryl Wimberley (paperback, 6 x 9, 320 pages, $19.95)

Please write to request our free catalog. To order by credit card, call 1-800-491-2808 or send a check or money order to Allworth Press, 10 East 23rd Street, Suite 510, New York, NY 10010. Include $5 for shipping and handling for the first book ordered and $1 for each additional book. Ten dollars plus $1 for each additional book if ordering from Canada. New York State residents must add sales tax.

To see our complete catalog on the World Wide Web, or to order online, you can find us at *www.allworth.com*.